Wild Jewellery

materials | techniques | inspiration

SARAH DREW

PHOTOGRAPHY BY TOM BARKER

jacqui small

First published in 2012 by
Jacqui Small LLP
An imprint of Aurum Press
7 Greenland Street
London NW1 0ND

PUBLISHER: Sarah Bloxham
PROJECT EDITOR: Samantha Warrington
DESIGNER: Rosamund Saunders
EDITOR: Anna Southgate
PRODUCTION: Rohana Yusof

ISBN: 978 1 906417 796

A catalogue record for this book is available
from the British Library.

2014 2013 2012
10 9 8 7 6 5 4 3 2 1

Printed in China

Contents

Introduction

There are plenty of good reasons for seeking out found materials with which to make wild jewellery. One of the best reasons is that it is great fun. The whole idea of foraging for treasures is almost as enjoyable as the making and wearing of the pieces they become so much a part of. A good forage always involves a walk in the open air; it gives direction and purpose to a day out; and it starts the design process rolling right there and then. Almost without knowing it, as you look for interesting things to incorporate in your jewellery, you begin to plan what to use in order to show them at their best.

Another great reason for making jewellery using found materials is that every piece you create will have a personal link to a certain aspect of your day out; it may be a reminder of a unique setting, of dramatic weather or of a particular person you went with. A pebble found in the local park, given to you by one of your children, perhaps, will make you smile when it's made into a pendant you wear and touch everyday.

Good green credentials

It goes without saying that the lack of environmental impact of using found materials is another positive aspect of making jewellery in this way. There are many places where you can find fallen twigs and wood or washed-up pebbles and shells to use in your designs. Even discarded bottles and papers can be put to innovative use to beautiful effect. That this can be done without costing anything to produce in financial or environmental terms is an added bonus. In fact, if you're collecting weathered plastic from the beach or ring pulls from the park you're contributing to a cleaner environment – making it more pleasant for the local community – and are recycling materials immediately into something that can be used again and loved.

A WORD ON FORAGING

It goes without saying that you must always take care when out looking. Make sure that you plan a day out at the beach, in the woods or in the city carefully, and within daylight hours. As long as you follow a number of practical guidelines, your day out will be fun.

◼ Check your location carefully: Make sure that all the places you intend to visit are public and not private property and avoid the temptation to stray. A penalty will almost certainly apply for trespassing.

◼ Dress appropriately: Keep an eye on the weather forecast and make sure you have the correct clothing for sunshine or rain, along with sunscreen, hats, gloves and so on. The locations in this book have varied terrains. You are likely to be out walking for several hours, often on uneven ground, so wear sturdy, waterproof footwear. If you intend to collect a lot of treasures, take suitable containers for carrying them and do not overfill them.

◼ Beware of danger spots: Take a map of your location with you and make sure you know where you are at all times. Stick to designated paths at the beach and in woodland areas. Be wary of turning tides, steep rock faces, disued quarries, woodland streams, native wildlife, traffic spots and the like.

◼ Respect the environment: Be sure not to cause any damage. If taking cuttings from plants, do so sparingly, using the appropriate tools. Observe any preservations that there might be on taking specimens from specific plants or pebbles and shells from protected beaches. Don't forage private gardens. Do not leave litter.

◼ Stay safe: Make sure that someone knows where you are going, especially if you are foraging alone. Always take a mobile phone, cash and provisions.

Besides all the feel-good fresh air and green virtue, found materials look fun and funky and are lovely to work with: a piece of opaque sea glass looks every bit as good as a chunk of aquamarine; woodland bones can be shaped and etched like ivory; twigs are pretty and free; and quirky finds from the streets around your house can star in sleek modern designs. Furthermore, in the current economic climate, if you can find interesting materials in abundance to work with, your creativity needn't be stunted by lack of funds. Combine natural, found things with old bits of remade silver and gold and you'll get truly unique pieces that you don't have to economize on: you can make a massive statement necklace out of sea plastic – it costs nothing. Slice the twigs as thick as you like, there are plenty and the trees need pruning. Pick up that plastic bag from the park and make it into a modern chain... the list goes on and fits in with the things you find in the normal environs of your own life.

Using this book

Wild Jewellery is all about creating pieces that get people talking. You might have tried making jewellery before, but previous experience is not essential. You can use the techniques to develop your own ideas – and that's the beauty of this book.

Take almost any technique presented in the following chapters and you can use it to make many different types of jewellery – pendants, earrings, bracelets, tiaras, rings, necklaces – you name it. You simply adapt the technique to your design and whatever you have to hand after a good forage. Better still, although the techniques are arranged according to the places you might visit – the beach, a woodland, in town – many of them are suitable for a wider range of finds. A crocheted stone setting works as well with a park pebble as it does for sea glass; you can crochet plastic bags as easily as you can crochet sea string; and you can set woodland moss in resin in exactly the same way as you would city-park grass. Demonstrated with step-by-step photography, the techniques are very easy to follow. There are Quick Size Guides to help adapt a technique to specific designs, and Finishing Tips to show the kinds of pieces you can make. Using found treasures is often about experimenting with new materials, so there is bound to be an element of trial and error, but that is what is so appealing about making wild jewellery. The main thing is to get yourself out and about, find some more good, free treasures and get making!

Sarah Drew

Tools and materials

The tools and materials that you need to work with the techniques in this book are listed on these pages. There is no need to invest in brand new tools when starting out – most of them can be bought second-hand, either on eBay or at a local market. Some good-value sites are listed at the back of the book for sourcing wire, beads and findings (see Useful addresses, pages 142–43).

MATERIALS The techniques use various threading materials, such as silver wire, tigertail and cotton cord. Silver-plated copper core wire is popular and easy to use: 0.4-mm, 0.6-mm and 0.8-mm gauges are good sizes. Sterling-silver wire is used for some projects involving soldering. Other techniques include setting finds in polymer clay or epoxy resin, and inlaying silver sheet or tubing. As well as having a good range of findings (clasps, jump rings, extender chain), it is worth looking out for lengths of pretty ribbon to use for finishing your pieces.

BASIC KIT You'll need a handful of basic tools, many of which you may have in the house already – scissors, PVA glue, hammer. You will also need wire cutters and snips for cutting wire and sheet metal; flat-nosed pliers, a mandrel and wooden dowelling for manipulating wire; and a hammer and flat plate for texturing metal. For crocheting, different size hooks create different effects, so invest in several. Recommended sizes include 1.2mm for fine silver wire, 2–3mm for sea string and 5mm for plastic bags.

SOLDERING EQUIPMENT Soldering is a way of joining two metal elements using solder and heat. Solder is a metal 'glue' with a lower melting point than that of the metals. Borax, a liquid flux, is applied and allows the solder to spread and fill the join. Pieces can become firestained during the process, and are soaked in a pickle solution for a couple minutes (using plastic tweezers). Goggles must be worn for soldering. This process is not suitable for copper wire.

MULTI-TOOL Various techniques can be speeded up using a small power tool with a number of useful attachments. Several of them, such as drill bits and sanding attachments come with the tool.

SAWING AND SANDING A handsaw is best for wood and twigs, while a junior hacksaw is good for bone, plastics and metal. Sanding and polishing can be done by hand. Various grades of sandpaper are recommended for wood – from coarse to fine – a needle file, wire wool and emery paper are best on metals, plastics and resin.

BASIC KIT (top left) snips; dowelling; craft knife; flat-nosed pliers; wire cutters; hammer; mandrel; ruler, glue; small, sharp scissors; household scissors; pencil; bench clamp

SOLDERING (top right) pickle; solder; fireproof brick; paintbrush; borax cone; small soldering torch

MULTI-TOOL (bottom left) power tool with drill bits; burrs; sanding, cutting and polishing attachments

SAWING AND SANDING (bottom right) sandpaper; emery board; needle file; hacksaw; handsaw; bench peg

Finds beside the seaside

With its happy-holiday associations and warm sunshine colours, the seaside is awash with inspiration for making all kinds of wild jewellery. The techniques selected for this chapter make the most of the materials you are likely to find, whether it be a handful of bright sea-plastic fragments, a delicate piece of smooth-worn glass or a mass of hardy nylon netting. With innovative wirework and threading techniques, as well as instruction on crocheting, riveting and silver-chain making, you can use the methods shown on their own or several in combination with each other to create a truly diverse collection of necklaces, rings, brooches, bracelets, earrings and more.

A beach forage

Anyone who enjoys a day out by the sea would agree that it is impossible to resist the temptation to forage! Meander along the sand or loll around on a beach towel, and you will naturally, probably quite absent-mindedly, be drawn to things that you want to pick up for a closer look – intricate-looking seashells, smooth-to-the-touch pebbles, art-collage bits of sea plastic, sun-warmed driftwood and crazy, colourful string. Almost everyone leaves the beach with a little collection of seaside finds stored away.

By collecting these quirky pieces of found treasure and making them into something you can wear, you can bring to mind the pleasure of a day at the beach for many months and years to come. Touch a piece and you instantly remember how and where you found it, who you were with – even what the weather was like! What better way to create innovative pieces of wild jewellery that are unique and personal to you?

Where to look

In practical terms, some beaches have more of one type of treasure than others. For starters, yours may be a sandy beach with wild grasses or pebbly with plenty of seaweed. Even then, the range of colours of the natural finds will vary tremendously. When it comes to the man-made bits and pieces that wash ashore, you may find a great supply of sea glass and little driftwood twigs, or perhaps masses of worn-down sea plastic and string. It really depends on the tide and, to some extent, on the industrial use of the sea in that area. You'll get to know what's good to find at your local beaches or holiday destinations just by looking. And, unless you really know the beach, it is best to go with an open mind as to what might crop up. It can be very frustrating foraging for, say, green sea glass on a beach where there isn't much at all, particularly if you happen to be passing over dozens of cool pebbles or shells in the process.

In general, the best place to find smaller bits of flotsam and jetsam is along, or near, the tide line. Pieces tend to get washed up with the incoming waves and left along the line at which the water recedes having lost its energy. You may also find odd treasures partially buried in soft sand, nestled among the pebbles or tangled up in seaweed. For bigger pieces of driftwood, sea plastic and sea string, have a dig around any rocks at the shoreline, as these sometimes offer some promising catches. Always collect pieces that appeal to you in size, shape and colour, even if you don't have an immediate plan for incorporating them into a piece of your jewellery – there is no doubt that you will find uses for them one day. In fact, there is a great deal to be said for amassing and storing a good selection of beach treasures so that you have plenty to choose from the moment inspiration strikes!

Sea string, glass and plastic

Beaches are always dotted with the colourful remnants of man-made materials – plastics, glass fragments and shreds of netting that have come adrift during fishing expeditions. They find their way ashore, washed and worn by the sea. Some plastics keep their bright colours, while others fade in the sunlight. Pieces of glass become soft, smooth-edged and opaque, and fishing nets work their way into rock crevices or find themselves tangled up with the seaweed. All of these items make perfect materials with which to make wild jewellery. Not only do they bring a range of exciting shapes and colours to your designs, but they are durable and easy to work with.

Sea plastic
Collect pieces of brightly coloured sea plastic that have been warped by the salt water, or that have texture. They are great for using as charms on handmade silver chains (see pages 48–51) or crocheted into sea string (see pages 38–39).

Sea glass
Sea glass can be found in a range of gorgeous, soft, sea-type colours – opaque whites, greens and blues. Their shapes vary and they are especially attractive when they have smooth edges that have been worn down by the sea.

Statement pieces
Larger beach finds, such as this lovely strip of perforated sea plastic, make striking focal pieces. Use them for mounting other beach treasures (see pages 32–33) or riveting together with other found objects (see pages 42–43).

Fishing string
Nylon fishing string is perfect for unravelling into separate strands and using for threading (see pages 21–23), or working into a focal necklace by crocheting different pieces together (see pages 34–37).

Matted string
Bundles of soft, tangled sea string find their way onto the beach. Left intact it can be crocheted into for a dramatic ring or chunky, soft cuff, and embellished using small beads or shells (see pages 36–37).

Braided rope
Rope like this can be found in single lengths or as part of a fishing net. Either unravel it for single strands or plait longer pieces together to make chunky cuffs and statement necklaces.

Driftwood, pebbles and shells

Some beaches are rich in pieces of driftwood, pale and washed-out after days, even weeks at sea, its smooth, tactile surfaces almost warm to the touch. Elsewhere, a pebble beach offers thousands of small stones in earthy tones, rounded or flattened by years of being pounded by the waves. Seashells vary from coast to coast. You'll find smooth, blue-black mussel shells with pearly insides; broad, fan-shaped scallop halves with radiating ribs; gnarly, prehistoric-looking oysters; ridged cockles and rock-pool winkles – inspiration enough for all manner of fantastic wild jewellery creations!

Flat driftwood
Irregular, and often angular in shape, driftwood fragments make great bases on which to mount other found objects and beads (see pages 32–33). Look out for interesting forms, sea-worn painted pieces and bits of plywood.

Patterned pebbles
Many pebbles have contrasting strata or pretty speckling and you can design pieces of jewellery that make a feature of this. They often look better 'wet', which you can mimic with a coat of varnish.

Large shells
These work well as focal pieces. If they don't already have holes you can drill them carefully with a sharp drill bit and thread them into your design. Make a silver-wire crochet setting to frame them before incorporating them into a design (see pages 40–41).

Flat pebbles
Small, flat pebbles are brilliant for wire-wrapping and crocheting into designs (see pages 28–29 and 38–39). Even a relatively plain-coloured stone will look lovely embellished in this way. Try snapping a pebble with your fingers to test its strength.

Driftwood twigs
Small pieces of driftwood are lovely to use in jewellery, especially as strong, linear connecting parts in necklaces. Some pieces may be crumbly, but you can strengthen drilled holes with glue before use.

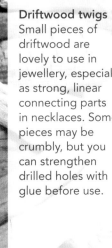

Holey shells
As well as looking pretty, seashells also sound lovely as jewellery, making a tinkling sound every time you move! Look out for pieces with natural holes from erosion, as they can be a bit tricky to drill.

Storage and preparation

It is actually surprisingly satisfying organizing and storing your beach treasures. Sort your pieces into different categories, grouping them by size, material or colour, depending on how you like to work. This will make it easier for you to find exactly what you want when creating a design. You can use almost anything for storage: old shoeboxes, empty biscuit tins, ice-cream tubs or any sturdy, good-sized containers will do.

Preparing pieces for use amounts to a handful of straightforward techniques – drilling, filing, sanding, unravelling – and these are quickly and easily carried out without any special equipment. However, in most cases, it is best to wait until you know how you want to use a piece before making any such preparations, as so much depends on the design of the piece of jewellery that you intend to use it for.

Cleaning beach finds

Most materials will have been bashed about by the sea, and so will already be scrubbed quite clean. You might find the odd piece of oil-stained sea string or plastic, which you can soak in warm water with washing-up liquid, before scrubbing clean with an old toothbrush.

Filing pieces

There will be times when you find a piece of sea plastic or driftwood that has rougher edges than you would like. Use a file to work backwards and forwards over the rough bit to make it smooth. Go carefully to avoid splitting the wood or cracking the plastic.

Sanding pieces

Even after filing, a piece of sea plastic or driftwood can still have a rougher finish than is required, especially for pieces that will be worn close to the skin. To rectify this, finish off by rubbing the surface with sandpaper for a lovely, smooth, tactile finish.

Drilling sea plastic

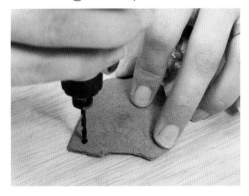

Sea plastic is easy to drill, but can be brittle and may crack. Having a sharp drill bit helps to make a neater hole, and a fast speed prevents the drill bit getting stuck. Wear goggles and turn the drill off to remove any plastic that gets stuck to the drill bit as you work.

Drilling driftwood

Allow driftwood to dry thoroughly before drilling, and take care that a piece does not split or crumble as you work. Go slowly with the drill and take your time. Rest the driftwood on an old piece of plywood to protect your work surface as you do this.

Unravelling sea string

Give a piece of sea string, rope or net a quick shake to remove loose sand and allow to dry. To unravel it, work from one end of a braid and use your fingers to comb through it. Extremely matted pieces of sea string can be used as they are for a less polished look.

Threading with sea string

This is a simple method for using sea string to make multi-stranded pieces in no time. What better way to incorporate beads, shells and other beach treasures into your jewellery designs? Experiment with different effects simply by varying the number and colour of strands and the type or density of the pieces you thread on them. Perhaps most satisfying of all, this is a technique you can work on at the beach, using whatever pieces of string you happen to find on the day.

TOOLS AND MATERIALS:
• fine strands of sea string, already unravelled (see page 21)
• your choice of beads and shells in different shapes and sizes
• scissors

QUICK SIZE GUIDE

Multi-stranded pieces work best with 15–20 strands. For a necklace, they need to be about 50cm long. For a bracelet, 30cm should be fine, and you could make dangly earrings with lengths of about 10cm.

1 Begin by cutting the number of strands of sea string that you would like to use. Stretch them out to their full length to make sure they are all long enough for what you are making.

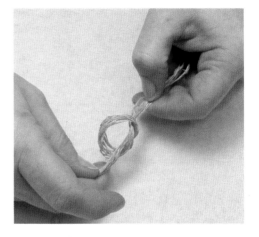

2 With all of the strands measuring roughly the same length (they do not have to be exact), start the piece by tying a tight overhand knot at one end to hold them all together securely.

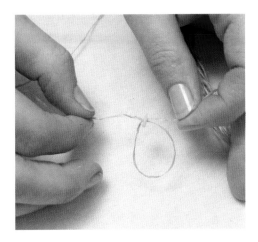

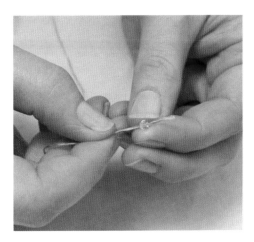

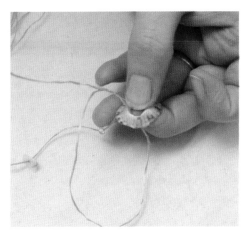

3 Working on one strand at a time, thread your beads and shells onto the sea string at random, depending on the look you are after. If you want a bead or shell to stay in a specific spot, loop the string back into it and out again, like making a stitch when sewing.

4 It can be effective to have some of the pieces moving about on the string, in which case skip the loop step every so often. This is a particularly useful trick for using tiny beads or shells where the hole is too small to thread the string through twice.

5 Consider the effect of the decorative pieces that you thread on. For a uniform look, select beads and shells of a similar size, tone and colour. To break the rhythm every so often, thread on a larger shell or bead, or one of a contrasting shape, texture or colour.

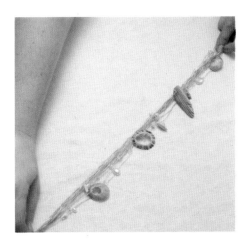

FINISHING TIP

As you complete each strand, it is a good idea to finish off with a bead or shell that is tied in place with a loop stitch, as a loose bead would simply fall off when you start to work on the other strands.

6 Thread some of the other strands in the same way – the design is up to you. It works well to use a different number of beads and shells on each one for that random, 'caught-in-the-net' look.

7 Now tie a second overhand knot at the other end to hold all the strands together securely. You are ready to attach a fastening to the piece (see Binding a fastening, page 47).

Threading with tigertail

Made up of several very thin strands of steel wire in a plastic coating, tigertail is strong and durable – ideal for connecting heavier beach finds to make a striking focal piece, or for weaving beaded strands together for a more chunky look (see page 26). You can also create striking static designs, using crimp beads to fix beads and beach finds in position on the wire.

TOOLS AND MATERIALS:
- selection of beach finds, pre-drilled (see page 21) or wire-wrapped (see page 28)
- selection of beads
- tigertail
- crimp beads
- wire cutters
- flat-nosed pliers

QUICK SIZE GUIDE

Think about the design of a piece – a tight-fitting bracelet or a low-hanging necklace – and cut the desired length of tigertail, plus an extra 10cm for the endings. For a focal arrangement, lay the pieces out on a work surface and measure their collective length or depth and add the 10cm.

Basic threading

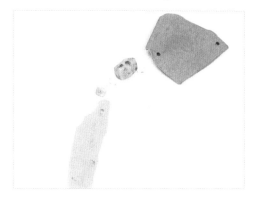

1 Decide on the arrangement of the pieces you want to thread. In this case a couple pieces of sea plastic alternate with some eclectic beads to create an interesting focal piece for a necklace. For pieces of sea plastic or driftwood, make sure that any holes are drilled quite close to the edge.

2 Cut a length of tigertail and thread two crimp beads on at one end. Thread the same end of the tigertail through the first beach find and then back through the crimp beads. Squeeze the crimp beads using flat-nosed pliers.

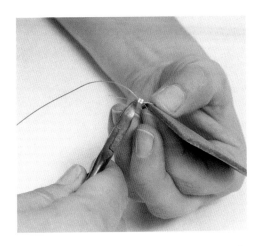

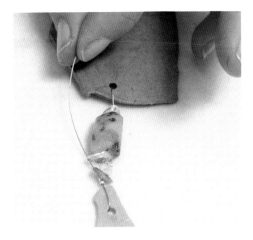

3 Use wire cutters to trim the short end of the tigertail close to the crimp bead. This completes one end of the focal piece, and is strong enough to take the weight of a number of additional pieces depending on your design.

4 Continue to thread your carefully selected pieces onto the tigertail. Here, a small glass bead is threaded on, followed by a chunk of polished granite. Together they complement the size and colour of the larger bits of sea plastic.

5 Before threading on the final piece, first slip two crimp beads onto the tigertail. Repeat step 1 to thread the wire through the last piece and back through the crimp beads, before squeezing with flat-nosed pliers.

Static threading

1 Decide on the position of a bead or shell on the wire and secure in place by squeezing flat a crimp bead either side of it. Leave the required space on the tigertail, before threading on another crimp-bead/shell-crimp arrangement.

2 Thread wire-wrapped pieces in the same way, passing the tigertail through the back of the wire-wrapping before flattening the second crimp bead. Think about the spaces you'd like between each feature bead or piece of sea glass.

3 With the design finished, you can thread a couple of crimp beads onto each end of the tigertail, followed by a clasp. Loop the wire back through the crimp beads before flattening them to complete the fastening.

finds beside the seaside

Weaving tigertail

1 Start by cutting a length of tigertail that is double the measurement you need. Thread both cut ends through two crimp beads, until you have a short loop at the bottom, and flatten the crimp beads to secure.

2 This design combines a number of small beads with larger pieces of wire-wrapped sea glass. Begin by threading around three to five beads onto each of the strands of tigertail. You can string them in any order you like.

3 With both strands complete to this point, weave the strands with one another once or twice then thread both strands through just one, slightly larger bead. This will gather the smaller beads together.

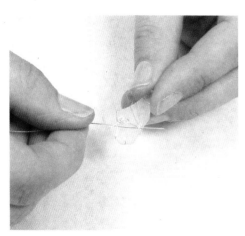

4 If using wire-wrapped pieces, thread them onto both strands in the same way, and push them right up to the woven beads. Because of their size, they will rest on top of the beads to create a more chunky effect.

5 Continue to thread and weave to complete the design. To finish, thread both ends of the tigertail through two crimp beads, add a clasp and thread the wires back through the crimp beads before flattening them.

FINISHING TIPS

This weaving technique effectively increases the strength of a tigertail piece, which means you can create much stronger, more chunky designs. It is particularly suitable for bracelets, which tend to be shorter and so less weighty with more pieces piled on in this fashion. If using the technique for a necklace, consider weaving for just a short section, rather than the whole length, to prevent a piece become too heavy to wear.

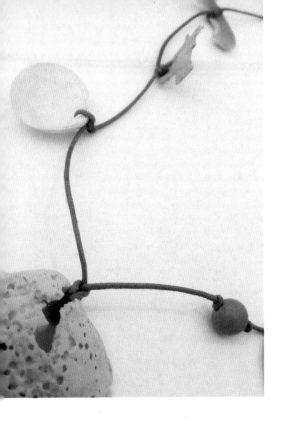

Knotting

Traditionally used for stringing pearls, knotting is a charming, yet straightforward way of making a necklace or bracelet with spaced-out beach finds and beads. Using a thin leather thong or cotton cord, this technique never fails to achieve a natural look. This is a jewellery-making method that you can easily do while spending a day at the beach.

TOOLS AND MATERIALS:
- selection of beach finds
- selection of small beads
- green cotton cord
- scissors

QUICK SIZE GUIDE

Knots use up a surprisingly large amount of cord, so it is a good idea to start with a length of cord that is at least twice the desired length of your finished piece.

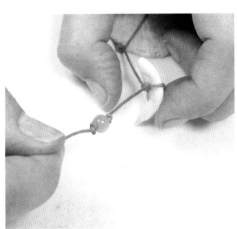

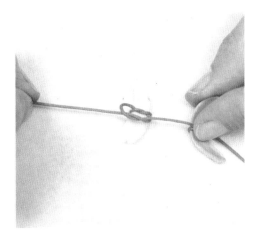

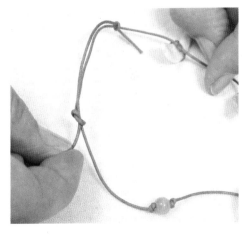

1 This technique really suits the random look demonstrated here. Simply decide where you want the first bead or shell to sit. Tie an overhand knot in the cord, thread the bead or shell on and then tie a second overhand knot right next to it.

2 To add a shell that doesn't have a natural hole, tie a firm knot around it and then tie a second knot on top of that. If the shell slips about, add a blob of glue. Continue to knot pieces on the cord in the same way.

3 Make a slip-knot fastening. Tie a loose knot in each end of the cord and thread the right-hand cord through the left-hand knot and vice versa. Place the piece over the head or wrist and pull on the cord ends to adjust the size.

Wire-wrapping

Some found materials, such as pebbles and pieces of sea glass, are quite hard to drill. A good way to overcome this is to wrap them in wire. It is aesthetically pleasing, but also offers a means by which you can incorporate them into your jewellery. There is no fixed way to do this and you are bound to develop a style of your own with practice. To use wire-wrapped pieces, thread silver wire or tigertail through the back of the wire-wrapping in the same way as you would thread it through a bead.

Wrapping a piece

TOOLS AND MATERIALS:
- flat pebble or sea glass
- 0.6-mm gauge silver wire
- wire cutters
- flat-nosed pliers

QUICK SIZE GUIDE

For an average size pebble or piece of sea glass (say the size of your thumb), approximately 40cm of wire should be ample. Don't go crazy selecting pebbles much larger than this, as they can make a finished piece of jewellery very heavy to wear!

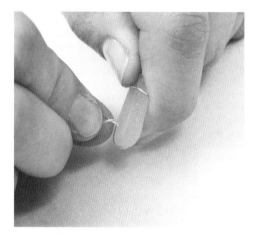

1 Take one end of the silver wire and wind it around the sea glass or pebble three times horizontally. Now wrap the piece three times vertically. You do not have to be wholly accurate, so long as the wrapping is taut enough not to let the piece fall out.

2 Now twist the two lengths of wire together in the centre of the pebble or piece. It is important that you keep the tension of the wire as you do this. Snip off any excess at the ends of the silver wire using wire cutters. The piece is now ready to use.

wild jewellery

Joining wire-wrapped pieces

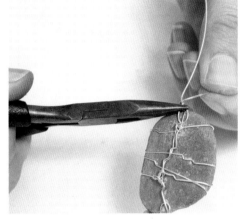

1 Cut a length of silver wire that is as long as your design, plus an extra 2cm at each end. Thread one end of the wire through the wire-wrapping at one end of your first piece.

2 Pull the end of silver wire through by 1cm and twist the short end around the long end to secure the loop you have just made. It may help to use flat-nosed pliers to get it tight.

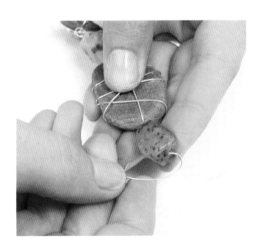

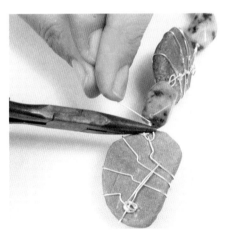

3 Continue to thread pieces onto the wire according to your design. Here, wire-wrapped pebbles alternate with polished granite beads. Simply thread each piece onto the wire in turn and push it close to the one it follows.

4 When it comes to using your last wire-wrapped piece, thread the silver wire through the wrapping at one end of the piece. Pull it through and wrap it around itself to complete the loop, before trimming any excess.

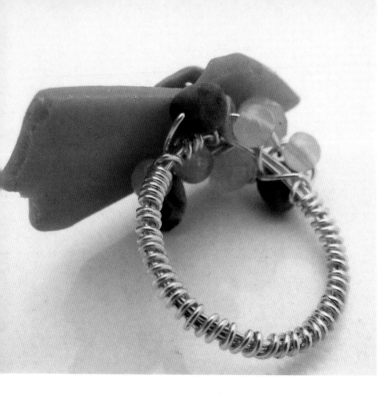

Making a wire ring shank

This is a great way of creating a simple wire base on which to showcase a pretty shell, a bright piece of sea plastic or a motley crew of leftover beads. The technique is straightforward and inexpensive, and can be achieved without the use of special tools. Even better, if you take a few items to the beach with you, you can make it right there with anything you spot on the day.

TOOLS AND MATERIALS:
- 0.6-mm gauge silver wire
- wire cutters
- flat-nosed pliers

QUICK SIZE GUIDE

You should be able to make a wire ring shank using 100cm of wire. If you don't have enough to go all the way around the shank, cut an additional 20cm of wire, wrap it on the shank where you finished up and continue binding the piece in the same way.

1 Take your length of silver wire and, starting about 12cm from one end, wrap the wire round the finger you'd like to the wear ring on three times: not too tight. Pull the wire off your finger, holding the loops together.

2 Bind the short end of the wire tightly around the three loops you've made, three or four times to secure them and snip off this tail end. If you are going to decorate your ring this is the time to do it (see Finishing Tips, opposite).

wild jewellery

3 To finish the ring shank, simply wrap the remaining length of wire all the way around the loops you made in step 1. Keep the wire tight so that you bind the shank snugly. This makes the ring stronger and smoother to wear.

4 Once you have reached all the way round to the other side of the shank, simply snip off any excess wire using wire cutters. Use flat-nosed pliers to squeeze the cut end inwards, flush to the ring, for a neat finish.

5 Try the ring on. If it feels loose you can make it smaller by holding the back of the ring shank firmly with your pliers and twisting your wrist by 90 degrees. This makes a small, Z-shaped kink in the shank for a slightly tighter fit.

FINISHING TIPS

 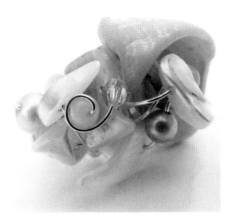 

The natural tones of the orange-brown sea plastic and polished malachite stone have been offset using bright primary-coloured beads.

Here, the soft, matt colours of a sea-worn shell are mirrored in the tones and textures of the beads with which it has been mounted.

This all-blue piece uses the wire-mounting techniques on pages 32–33 to incorporate two pieces of sea plastic into the design.

Mounting with wire

Wire is an ideal medium for holding larger pieces together that can't be heated. Here it has been used to create a quirky, sculptural form, while mounting it directly onto a base – in this case a found metal headband. You could just as easily attach a dramatic design to a torque or cuff base in the same way.

TOOLS AND MATERIALS:
- 0.6-mm gauge silver wire
- jewellery base – a headband, torque or cuff frame
- selection of large beach finds and found materials, pre-drilled where necessary (see page 21)
- 30–50 beads
- wire cutters
- flat-nosed pliers

QUICK SIZE GUIDE

The length of wire you need depends on the complexity of the piece. This example uses approximately 200cm. A torque will use a similar length, while a cuff may be half this.

1 Cut a suitable length of silver wire. Wrap one end securely to your base. Thread the wire up through the first hole in your largest decorative find, in this case a piece of driftwood, and wrap it around the base again to secure it beneath the wood.

2 This largest piece is likely to have more than one hole for securing to the mount. In this case take the wire along the base and attach the opposite end in the same way, by threading the wire through it and wrapping it back around the base.

3 Soften the join between a mounted piece and the base by threading on a bead. Push it right down and make a half-wrap around it from one hole to the other. Wrap the wire round the base once to secure it. Use a couple beads to create a more random look.

4 To attach a second focal piece – in this case glass taken from a vintage dress clip – use the wire to mount it in position. Here, the driftwood has a third hole for this purpose. Soften the join, if necessary (see step 3), using pliers to help pull the wire through firmly.

There may be times when you find a piece of jewellery or part of a child's toy on the beach. Perhaps something small that got overlooked when packing up after a swim, or a trinket that has washed up having found itself in the sea. These items make quirky focal pieces and often have the same sea-worn look about them. Most pieces are easy to incorporate into a design using wire-mounting techniques. You simply need to remove unwanted parts and make sure there is a hole for threading onto the wire. In the example above, a large plastic disc from a game makes the perfect base for wire-mounting a shell and beads.

5 You could finish here, but if you'd like to add a bit more decoration, thread beads onto the wire, pushing them into place where you'd like them to sit on the piece. To hold a bead in place, use the wire to make a half-wrap around it, from one hole to the other.

6 Leave a gap every so often, if you like, making a nice soft, sculptural curve in the wire with your thumb and finger instead. Tether the wire to the base at intervals if you can, to prevent the finished design from jumping about or working loose when worn.

Crocheting with sea string

Crocheting with sea string is an easy technique to master and provides you with a versatile base on which to develop a host of jewellery designs. All you need is a crochet hook and unravelled strands of sea string. You don't even have to be at home to do this, which makes it an ideal technique for working on the beach, incorporating a handful of treasures found on the day – pretty shells with natural holes and beads, for example.

TOOLS AND MATERIALS:
• strands of sea string, unravelled (see page 21)
• crochet hook
• scissors

QUICK SIZE GUIDE

Measure your neck or wrist using a piece of string and crochet the first row. A length of 40cm is good for a chunky necklace, and 20cm for a bracelet or cuff. You could also make a statement brooch using an approximate length of 10–15cm. The number of rows you choose to crochet is entirely up to you.

Crocheting in rows

1 Choose your string, combining colours if you like. Try to select strands of sea string that are equal to each other in length, as this will give you a much neater finish. Tie a knot at one end of the string to make a loop.

2 Push the crochet hook through the loop from front to back. Wrap the strands of string around the hook in an anticlockwise direction, above the first loop. Use the crochet hook to pull the string through to the front.

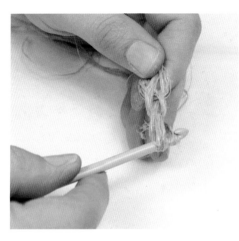

3 Hold the first loop firmly with one hand as you slide the crochet hook in the other hand towards you. Bring a short length of sea string through the first loop, keeping it on the crochet hook and fairly loose.

4 You will have made a second loop, linked to the loop you made in step 1. Now repeat steps 2 and 3 – wrapping the string anticlockwise around the hook, and pulling it through the loop – to make a third linked loop.

5 Continue in this way to create a length of linked loops that looks like a chain. Keep your work nice and loose so that you will be able to crochet the next row of linked loops onto the previous row, and so on, with ease.

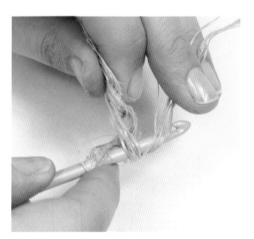

6 Once you have crocheted to your desired length, you need to 'turn the corner' to crochet the second row. Keeping the last link on the hook, push the hook into the next link along so that you have two loops on the hook.

7 Now wrap the sea string round the hook, still in an anticlockwise direction, and slide the string through both loops on the hook. This completes the first row, and starts the second, leaving you with just one link on the hook.

8 Continue to crochet the second row and repeat until you have the number of rows you need. To bind off, trim the remaining sea string to about 5cm and pull it through the last loop on the hook to fasten. Trim again to neaten.

finds beside the seaside

Crocheting frayed fragments

1 Start by selecting your fishing net fragments. You can pull several strands from a mass of found string – choose lengths that are particularly frayed.

2 Assemble several strands – at least ten – and follow steps 1 to 5 on pages 34–35 to make your desired length of linked chain-shaped loops.

3 You will quickly see how using frayed string creates a softer, more jumbled look. Using strands of varying lengths will add to this effect.

STATEMENT PIECES

Crocheting techniques offer a great opportunity to use those bits of fishermen's rope or netting that you see washed up by the sea. The colours can be bold, but also sea worn and make a perfect base for embellishing with all manner of wonderful sea finds. This is an effective method for making a strong statement necklace or a deep, dramatic cuff that rules an entire outfit (see the Finishing Tips on the opposite page for inspiration).

4 If you start to run short of netting, you can always add more. Simply knot a new fragment onto the end of the previous one, as necessary.

5 Crochet a second row (and any subsequent rows), following steps 6 to 8 on page 35 – the basic techniques for crochet remain the same.

wild jewellery

Crocheting in the round

1 Start with a crocheted length of sea string – one or two rows deep. Curve it round to form a rough circle and crochet the next stitch through the link at the opposite end to the one attached to your crochet hook.

2 If you want to make the piece a bit bigger, you can crochet a row around the outside of this circle. And you can fill in the middle of the circle in the same way, tying on a new length of sea string if you start to run short.

3 A piece like this could be used to make a brooch or a centrepiece for a hair embellishment. Because it is fairly small, it is best to keep any decorative elements to a minimum, as here with a wire-wrapped piece of sea glass.

FINISHING TIPS

The basic crochet techniques shown on pages 34–37 can be worked with fine wire, as shown in this chunky gold and sea glass necklace.

The foundation for this bright, large-scale seaside necklace is a crocheted base using three complementary colours of found sea string.

This elegant sea glass and pebble cuff combines crocheting with wire with the wire-wrapping techniques demonstrated on pages 28–29.

Embellishing a crocheted base

Large pieces of crocheted sea string or netting make the perfect base on which to build bold, colourful statement jewellery. One of the simplest ways to embellish them is by attaching decorative pieces using crochet. This is a versatile technique that allows you to incorporate a wide range of items into your design, either singly or as lengths of beaded wire.

TOOLS AND MATERIALS:
- crocheted sea-string base (see pages 34–37)
- 0.4-mm gauge silver wire
- 20–30 coordinating glass or stone beads
- selection of beach finds, wire-wrapped (see page 28) or pre-drilled (see page 21)
- wire cutters
- flat-nosed pliers
- crochet hook

WORKING A DESIGN

Before you start, lay your base flat on a work surface and loosely arrange your beach finds over the top, moving them around until you have a design that you like. This is particularly useful if you have a number of large pieces that might work better as one arrangement than another.

1 Start by cutting a length of silver wire and wrap one end around a top corner of your base – in this case the start of a large bib necklace. The length of wire does not need to be exact, because you can always add more as you work (see step 5). Around 150cm is a comfortable length to start with.

2 Thread your chosen beads onto the other end of the wire to make a row of about 15 to 20. Once complete, use flat-nosed pliers to make a little loop at the unwrapped end of the wire to stop the beads from falling off.

3 Now put the crochet hook through your base, from front to back, where you wrapped the wire in step 1. Take the wire anticlockwise around the hook and pull it through the sea string. This will have secured the wire to your base.

Single pieces

1 Single pieces may have pre-cast hooks through which you can attach them, or they may be pebbles and pieces of sea glass that you have selected specifically and pre-drilled or wire-wrapped. Either way, thread the wire you are working with through the back of the wirework or pre-cast hook.

4 Before you make the next crocheted link, push the first bead all the way up the wire so that it sits close to the sea-string base. Now you can wrap the wire around the crochet hook as you did before and pull it though the sea string. Making this link will hold the bead securely in place.

5 Work along your base, from one end to the other, securing the remaining beads in the same way. Continue until you are happy with the level of embellishment. Attach additional lengths of threaded wire as necessary, wrapping the end around a few strands of sea string before crocheting in place.

2 Crochet the piece to the sea-string base. Put your hook through the sea string, from front to back, and wrap the wire around the hook in an anticlockwise direction. Then, using the hook, pull the wire through the string. Attach additional single pieces to the base in the same way.

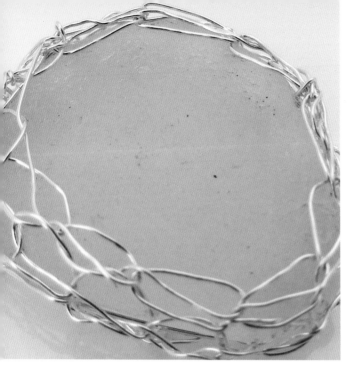

Crocheting a stone setting

This is a very pretty way of incorporating sea glass, found pottery and pebbles into jewellery designs without having to drill or solder a setting. The technique is a delicate one, so think carefully about the pieces you choose in terms of size and colour. Once you have a selection of lovely crocheted beach finds you could use them to make drops for earrings or a pendant, or alternate them with beads to make an elegant bracelet.

TOOLS AND MATERIALS:
- flat pebble or sea glass
- 0.6-mm gauge silver wire
- 0.3-mm gauge silver wire
- crochet hook
- flat-nosed pliers
- wire cutters

QUICK SIZE GUIDE

Avoid pieces that are too large – 3 x 2cm is really the biggest size recommended, for which 15cm of wire is long enough. You can use wider gauge wires (0.8 and 0.4 mm, for example) for the largest pieces. A 1.2-mm crochet hook was used here – the larger the hook, the more holey the effect.

1 Make a basic mount. Use the 0.6-mm gauge wire to form an outline of the pebble or sea glass and twist the two ends together using flat-nosed pliers. Make the shape a little smaller than the actual piece to stop it falling through.

2 Now take a length of 0.3-mm gauge wire and wrap one end tightly around your wire shape to attach it securely. Beside this, make a loop for the crochet hook to go through, by doubling the wire back on itself a little and twisting.

3 Put the crochet hook through the loop and crochet a link (see Crocheting sea string, pages 34–35). Bring the hook through the wire frame and crochet a second link. Now take the hook outside the frame and crochet a third link.

4 Bring the hook through the wire frame to make a fourth link. Continue your way around the frame in the same way, crocheting a link inside and outside alternately so that the frame is worked in a complete row of crochet links.

5 When you've gone all the way around the frame, simply work a second row into the first by crocheting into each link. In this way, you add to the first row, building up a crocheted frame in which to set the pebble or sea glass.

6 Having completed the second row, position the pebble or sea glass in the base and use your hands to squash the crocheted frame around its edges. It needs to be a reasonably tight fit to prevent the piece working loose.

7 Crochet a third row of the frame with the pebble or sea glass in place to cradle it. If it is a bit loose, drop a link now and then. You might need to crochet a fourth row, depending how deep your chosen piece is.

8 Once you are satisfied that the pebble or sea glass is securely held in place, return to your starting point and finish off by wrapping the end of the wire around the base a couple of times. Trim off any excess wire.

Riveting

This is a great cold-connection technique for joining materials together. The piece here shows the method being applied to sea plastic, but it is equally effective used on metal, wood, plastic and all sorts of found materials in a decorative, edgy kind of way. It helps to create a great look for quirky brooches, focal pieces for cuffs and statement necklaces like the one shown here.

TOOLS AND MATERIALS:
- two or more pieces of sea plastic or other found material
- silver tubing
- masking tape
- pencil
- multi-tool with drill bit
- junior hacksaw and spare blades
- ruler
- metal plate
- hammer
- large nail
- needle file
- fine silver chain
- superglue

1 Tape two or more pieces of plastic together using masking tape, and use a pencil to mark a number of riveting points. Carefully drill through all layers of the taped plastic, using the same diameter drill bit as your tubing.

2 Slot the silver tubing into one of the drilled holes, so that it goes right through from front to back. Use the pencil to measure the point on the tubing, at which you need to saw it. This mark needs to sit a little bit higher than the surface of the plastic.

3 Carefully cut the tubing to this length using a hacksaw. Measure the tubing for subsequent holes in the same way, or transfer the measurement from the first piece using a ruler. Insert one piece of cut tubing into each of the holes.

4 Make sure that each piece of tubing goes through all the layers of taped plastic. They may slot into place easily by hand. If not, place the taped plastic on to a metal plate and use a hammer to tap the tubing gently.

5 Slot the large nail into a piece of silver tubing (the nail should be only slightly larger than the tubing in diameter). Tap the nail very gently with the hammer, so that the edges of the tubing splay out a little.

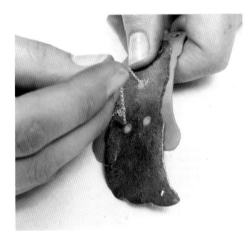

6 Repeat this process for all the tubing, then turn the whole piece of plastic over, keeping it on the metal plate. Continue to tap each piece of silver tubing gently. You want to flatten the tubing out slightly so that it doesn't move about in the hole.

7 Don't rush the hammering process or apply too much pressure, as the plastic can be brittle and might easily split or crack. Once the hammering is complete, use a needle file to file the metal tubing on both sides of the plastic for a smooth finish.

8 For a more 'punk' look, you can thread lengths of fine silver chain in and out of the rivets in a haphazard fashion. To secure the ends of the chain lengths, dab the back of the riveted sea plastic with a little blob of superglue. Allow to dry thoroughly.

Making pull-through earring chains

This is a simple technique for making chains that you can pull through your ears, and to which you can then attach all manner of beach-found earring drops (see page 46). You don't have to make sure that each earring is exactly the same length when you pull the chains through – a little asymmetry is all part of the appeal, particularly as you are likely to have mismatched drops.

1 Use wire cutters to cut two identical lengths of 0.8-mm gauge sterling silver wire for each pair of earrings. These will be used as the earrings posts, making it easier and more comfortable to thread the earrings through the ears.

TOOLS AND MATERIALS:

- 0.8-mm gauge sterling silver wire
- fine sterling-silver chain
- wire cutters
- fireproof brick
- silver solder
- metal plate
- hammer
- snips
- paintbrush
- borax cone
- goggles
- soldering torch
- flat-nosed pliers
- plastic tweezers
- pickle
- emery paper
- needle file

QUICK SIZE GUIDE

Obviously, you can make the pull-through earrings any length you like! This example uses 1.5-cm wires for the earring posts and 10cm for the chains.

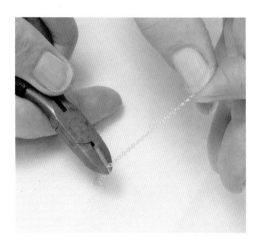

2 Cut two identical lengths of silver chain for the earring drops. Lay each chain out on your fireproof brick, a little way apart from each other and without any kinks. Place the earring posts on the brick, each one overlapping the last link of a piece of chain.

3 Prepare your solder. Place it on a metal plate and use a hammer to tap it gently a few times. This will make it more malleable. Use snips to trim off one piece of solder for each earring (about 1mm in size).

4 Use a paintbrush to wet the borax a little. You may already have the borax cone standing in a water, in which case you will have some solution to hand. Paint a little on to the end of each silver chain where it overlaps the wire post.

5 Still using the paintbrush, pick up a small piece of solder in turn and place it on the same join. Wearing goggles, light the torch and set it on a low flame. The hottest is the point of the blue inner flame: aim this at the wire.

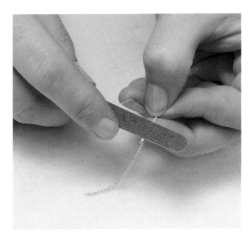

6 Move the flame quite quickly over the solder, so that all elements heat up evenly. The wire and chain are fine and the melting process happens quickly. The solder melts just after it starts to glow red. You should see the wire and chain settle as they fuse together.

7 Use flat-nosed pliers to pick up the fused pieces – they will be too hot to handle with your fingers. Cool them down under a tap or in a bowl of cold water. Place them in the pickle to remove the fire stain, dropping them in carefully or using plastic tweezers.

8 When they look bright silver, take the earrings out and rinse under water. Clean the surface with some emery paper and check each end of the wire posts. Use a needle file to smooth them where they appear sharp or bulky. Take care, as the soldered joint is fragile.

Attaching an earring drop

You can use this method to attach all manner of quirky beach finds to the pull-through earring chains on pages 44–45. You want to keep the drops small and light, so be careful when selecting them. Experiment with different shapes and colour, for mismatching earrings, or make a pair of near-identical earrings to complete a set.

TOOLS AND MATERIALS:
- Sea plastic, pre-drilled (see page 21), a shell or a pebble, wire-wrapped (see page 28)
- 8-cm length of 0.6-mm gauge silver wire
- pull-through earring chain
- flat-nosed pliers
- wire cutters

1 Thread the wire through a hole or loop in the drop. Pull the wire through a little and use the short end to secure the drop by making a closed wrapped loop. Thread the other end of the wire through the end of the earring chain.

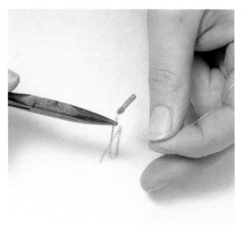

2 Use flat-nosed pliers to close the loop by wrapping the wire neatly back around itself. Make sure the wrapping is not too tight – the drop should have freedom of movement. Use wire cutters to trim off any extra wire.

MIX AND MATCH

Since no two beach finds can ever be truly identical, you have many choices when it comes to making and wearing a pair of earrings. You can take the time to find a close-matching pair – say two similar pieces of blue sea plastic. Or you can go all out for really odd-matching pairs, using a shell for one and a knot of sea string for the other. Be warned though. Wearing odd earrings causes a stir: people just have to tell you you've got different earrings in!

Binding a fastening

This is a quick and simple method for completing a bracelet or necklace made from sea string or fishing net. It is most suitable for lightweight pieces, where there is less likelihood of any pull from heavy embellishments, but you can always experiment to see what works best for you. This is a versatile technique that you could also use to bind the top of a dangly sea-string earring, ready for inserting an earring hook into. Once the wire-wrapping is complete, you simply trim any excess wire using wire cutters and push the cut end into the binding.

TOOLS AND MATERIALS:
- 0.6-mm gauge silver-plated wire fine silver chain
- hook clasp (see page 53)
- flat-nosed pliers
- wire cutters

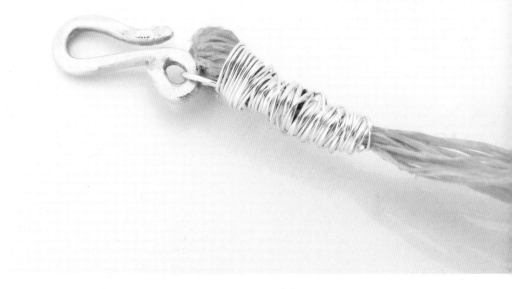

1 Bind one end of your piece tightly with a 50-cm length of silver wire. Start just before any knot, tying the strands together, and continue over the knot. Use pliers to gain some leverage so that your binding is really tight.

2 Three-quarters of the way along the knot, make a small loop in the wire and thread on the extender chain. Wrap the wire back on itself to close the loop. Repeat at the other end of the piece, attaching the hook clasp instead.

Silver-chain making

This technique uses 1.2-mm gauge wire to make a silver chain that is particularly suitable in terms of both scale and strength for holding heavy beach finds, such as pebbles and larger pieces of driftwood. You can also use 1.0, 1.5 and 2.0-mm gauge wire. Making your own silver chain allows you to have so much more control over how your jewellery looks. You can give pieces a much more original, lasting design, tailor-made to your ideas and vision.

TOOLS AND MATERIALS:
- 1.2-mm gauge silver wire
- piece of dowelling
- masking tape
- wooden bench peg
- junior hacksaw
- fireproof brick
- flat-nosed pliers
- borax cone
- silver solder
- metal plate
- hammer
- snips
- paintbrush
- goggles
- soldering torch
- plastic tweezers
- - pickle

Preparing the links

1 Start by wrapping a length of 1.2-mm gauge silver wire firmly around the piece of dowelling to make a tight coil. The length of the wire depends on how many links you intend to make, but 10–15cm is easily workable.

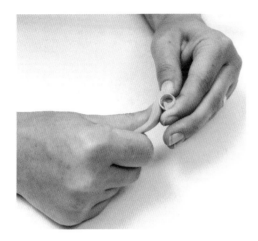

2 Carefully slip the coil off the piece of dowelling and wrap the entire length of it with masking tape. Place the wrapped coil in the 'V' of a wooden bench peg so that you have something firm to lean against when sawing.

3 Saw through the coil of wire gently, working in a vertical motion. Take your time so that you don't bend the circular coil out of shape. Saw blades break quite easily so make sure you have plenty of spares (they're not expensive).

4 Once you have sawn through the coil of silver wire, you need to peel off the masking tape. Go slowly and carefully. The sticky tape will pull the cut rings out of shape a little, and you want to avoid too much distortion.

5 Remove each silver link from the masking tape, one at a time, and place it on a fireproof brick ready for soldering. You can see here how each one has lost its circular shape just a little during the removal of the masking tape.

6 To make a length of silver chain, you need one unsoldered ring for every two soldered rings. Use flat-nosed pliers to make sure that the cut ends of each link are aligned as closely as possible with each other.

7 Before soldering the links, make sure you have all the equipment to hand. It is useful to have the borax cone standing in a small amount of water. This means that there is always a little borax solution in the dish, ready for use.

Soldering the links

1 Soldering the links strengthens the chain. Place the solder on a metal plate and hammer the end a little. This makes it easier to cut off small pieces.

2 Use snips to cut several 1-mm pieces of solder from the hammered end of the strip. You will need one piece of solder for each silver link.

3 Paint a little borax solution over the join of each ring. Still using the paintbrush, pick up a small piece of the solder and place it on top of the join.

4 Wear goggles as you heat up each ring (for the solder to melt and seal the join). Move the top of the blue flame over the ring in a circular movement. Both cut ends of the ring need to heat up at the same rate so that the solder melts and stretches between them.

5 When the solder has melted, use pliers to pick up each ring, cool under a tap or in a bowl of cold water then drop it in the pickle to remove the black fire stain. Don't let your pliers go into the pickle as they'll contaminate it and make all the silver turn pink.

6 Once shiny again, use tweezers to remove the soldered links from the pickle and wash and dry thoroughly. Place each on a metal plate and flatten with a hammer, giving it a slightly textured finish. (Do not hammer the unsoldered links yet.)

wild jewellery

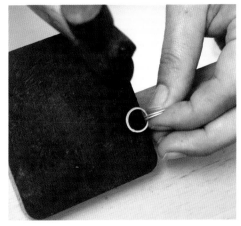

7 Now thread one of the unsoldered rings through two soldered rings to join the three together. The new, unsoldered ring will become the middle link of the three once soldered.

8 Follow steps 1 to 4 to solder this link. Cool the whole piece in water and then place it in the pickle. You can keep adding unsoldered links in this way until you have your desired length of chain.

9 Each time you solder a new link to the existing chain, remember to give it a hammered finish to match it with the others. This is best done as you add each new link, rather than at the end.

FINISHING TIPS

Silver links of two different sizes have been used to connect the quirky, misshapen pebbles in this robust-looking necklace.

A lucky charm bracelet with a twist, where each link of the silver chain has a colourful bead or quirky beach find attached to it.

This asymmetrical piece uses just three links of silver chain. They are attached to a piece of driftwood using an S-link (see page 52).

Making cold connections

These handy links and hooks come into their own when making jewellery using your beach treasures. They provide connections between materials that cannot be soldered together because they would either ignite or melt. You can use S-links to connect lengths of silver chain to pieces of driftwood and sea plastic, while the hooks offer a simple means of fastening a piece. Make a number of each at a time so that you always have some to hand.

Making an S-link

TOOLS AND MATERIALS:
- 6cm of 1.2-mm gauge silver wire per link/hook
- goggles
- soldering torch
- flat-nosed pliers
- plastic tweezers
- pickle
- hammer
- metal plate

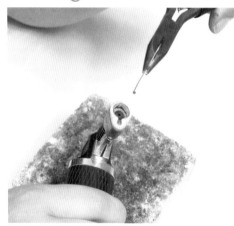

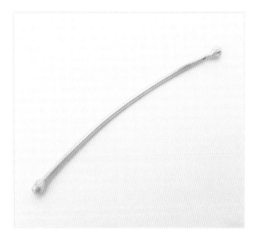

1 Wearing goggles, light the torch. Hold a length of wire in the pliers with one end over the hottest part of the blue flame. The wire has to be vertical to create neat balls so hold the torch on its side. A ball will form quickly.

2 Cool the wire in a bowl of water and repeat the process at the other end of the wire. Using plastic tweezers, put the wire in the pickle to clean it off. Prepare all the wires for a particular piece to this stage before proceeding.

3 When it comes to using an S-link, take one of your ready-made wires and thread one end through a hole on one of the pieces you wish to connect – in this case a piece of driftwood. Once through the hole, bend that end of the wire into a 'U'-shape using pliers.

4 Do not bend the wire too closely at this stage. The piece you are connecting needs to have some movement. Hammer each of the balls flat, if you wish. This gives a piece a neater look, but also serves to strengthen the silver a bit more.

5 Curve the other end of the wire in the opposite direction to make a loose 'S' shape. Thread your next piece onto this before tightening each curve to complete the link. Keep the curves broad enough so that there is still some freedom of movement.

Making a hook clasp

1 Take a 6-cm length of wire and use flat-nosed pliers to curl one end of the wire into a small, tight loop.

2 Use your thumb to curve the other end of the wire, then complete the 'shepherd's-hook' shape using pliers.

3 Place the wire on a metal plate and solder the loop you made in step 1 to secure it. Pickle the hook to finish.

Beach inspiration

Beach-found materials lend themselves to quirky, dramatic jewellery, allowing you to create completely original pieces. You can combine contrasting weathered brights for a cheerful summery look; moody grey slate and black plastic for a wintry stormy look, or aqua/green glassy colours for a more restful, ethereal feel. Combine textures and techniques for a truly eclectic look.

Blue sea-plastic pendant
A found brooch makes a bold focal piece. (See pages 32–33 for mounting with wire.)

Sculptural wirework
Bright sea plastic and a few choice beads are piled high for a contemporary ring.

Chunky crochet
Big, strong embellishments complement thick rope. (See pages 34–37 for crocheting.)

Subtle connections
Tigertail can link materials you can't solder. (See pages 24–26 for using tigertail.)

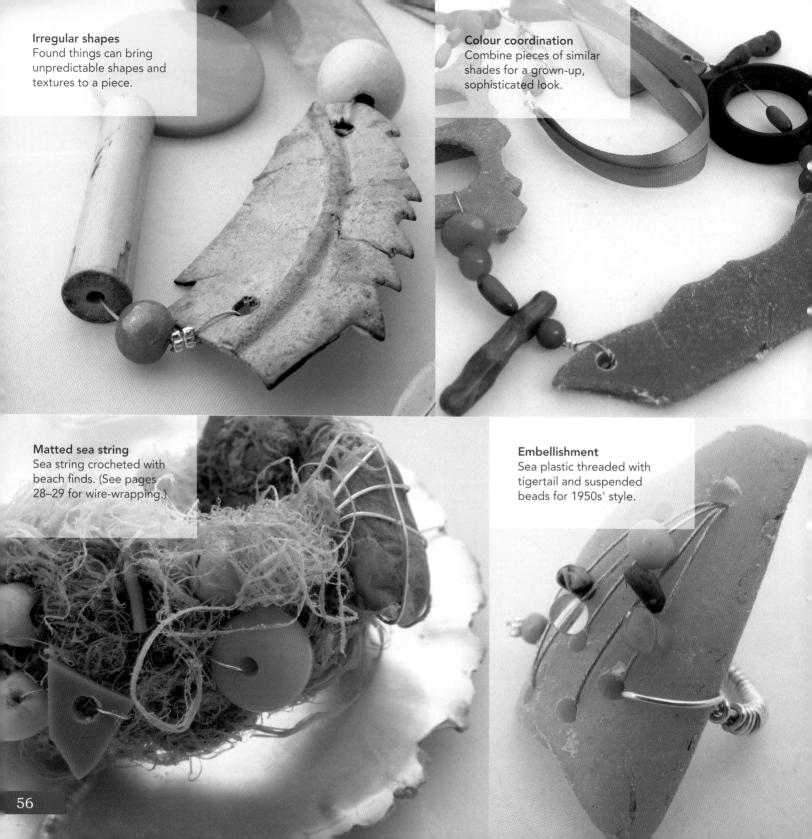

Irregular shapes
Found things can bring unpredictable shapes and textures to a piece.

Colour coordination
Combine pieces of similar shades for a grown-up, sophisticated look.

Matted sea string
Sea string crocheted with beach finds. (See pages 28–29 for wire-wrapping.)

Embellishment
Sea plastic threaded with tigertail and suspended beads for 1950s' style.

Jewellery by Adele Stanton

Working in Penzance, on the coast of Cornwall, Adele Stanton makes beautiful linked bracelets, necklaces and rings inspired by the sea. The name of her company, Hastha Kala, translates as 'art with your hands' and many of the components she uses to make her jewellery are handmade or hand-finished. They include silver soldered links, hammered silver feature pieces and handmade chain.

Characteristic of Adele's work are the sea-glass fragments that she mounts in sterling-silver rub-over settings. She polishes the glass in a stone tumbler to smooth the rough edges, then shapes the silver base of a setting to mirror that of the glass and uses thin bezel sheet to surround each piece. Adele uses beach china and pretty pebbles in the same way.

Top: Green sea glass set in sterling-silver rub-over mounts and joined with fine silver links.

Left: Delicate sea glass and hammered silver necklace. (See pages 48–51 for making silver chain.)

Into the woods

Some of the most magical, fairy-tale-inspired wild jewellery can be created using the natural materials that abound in a woodland setting. There are diverse treasures to discover here, each providing its own inspiration. Consider the ticklish feel of a downy feather or the hard, polished surface of a freshly fallen conker. Playing to the unique qualities of such finds, the techniques presented in this chapter include innovations like the 'rattle-ring' shank on which to cluster a bundle of seeds. There is also instruction on working with epoxy resin, on inlaying with silver and with sculpting delicate creations using free-form wirework.

A forage in the woods

Old clothes, boots, a coat with big pockets: that's all you need for a woodland forage. Whatever time of year you go, you'll find plenty of free, beautifully formed materials from which to make wonderful, wild, woodland jewellery: spring twigs, shiny conkers, lost feathers and perfectly formed acorn cups. Jewellers have been replicating these natural beauties in precious metals for centuries; why not use the real thing since it's available at your fingertips?

Some of the materials you find won't last forever and may even change colour or shape after harvesting, but this is part of the appeal of making woodland jewellery. It acts as a reminder as you wear your acorn earrings or your bent twig ring, that everything does change and life moves on. Not only that, but it allows you to go collecting again, make more things and experience something new. Of course, there are also materials and techniques that are free of such transience, and which result in pretty, solid pieces that last for a much longer time – those using silver inlay or epoxy resin, for example.

A good number of woodland offerings can be used as you find them – downy feathers, little fir cones, moss – while others need a little preparation, such as drilling, washing or stripping. Bone and larger pieces of wood require more work in order to achieve a truly satisfying finish. Yet this is by no means a negative aspect of working with these things. Instead, it shows that there is a tremendous resource of raw, sustainable materials available to you all year round and from which you can create anything you like, provided you invest in some basic tools and learn some useful techniques (see Tools and materials, pages 10–11, Storage and preparation, pages 66–67 and Working with wood, pages 68–69).

Fairy-tale inspiration

What you make with your woodland treasures depends on what you have in mind, but also on what you find! See what catches your eye as you stomp through muddy puddles or crunch through autumn leaves and fill your pockets with quirky, beautiful, natural wonders in vivid colours and with bold textures. More than any other outing, a forage in the woods evokes all sorts of childhood memories, loaded with an inescapable fairy-tale element borne from generations of storytelling. Embrace the wonder of it all, and you may be surprised to find echoes of this in the pieces you make.

Twigs, bark and seeds

One of the true pleasures of seeking woodland treasures for your wild jewellery lies in the many textures there are to be found. Lichen-covered twigs, rough-ridged bark and smooth, shiny chestnuts – each is tactile in its own way and lends itself to many applications when it comes to jewellery making. Many seeds are, of course, seasonal and an autumn forage can be most rewarding as you trek along woodland paths in search of horse chestnuts yet to open, acorns complete with their little cups attached and bunches of delicate sycamore wings. Harvest green twigs as well as those already dried out, as both have their uses.

Wood knots
Cutting into a wood knot is like opening a surprise gift: you really don't know what to expect, but you know it will be lovely. By using wood knots in your jewellery, you guarantee that each piece you make is absolutely unique in design.

Fir cones
There are plenty of little pine cone varieties around – perfect in both shape and size for making earrings and pendants, for layering up with beads or twisting with wire (see page 73). They are a very versatile natural material.

Twigs
Thin, green twigs can be bent and layered up to make necklaces, cuffs and tiaras (see pages 76–77). Thicker twigs, cut from trees, can be sliced into beads and decorated with paint or resin (see pages 86–88). Twigs taken from the woodland floor are often too brittle to use.

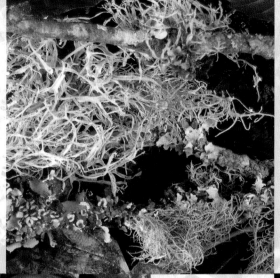

Bark
Bark is always useful as a base for decorating with other found objects. Long, flat pieces work well in multi-necklaces or inlaid with resin or metal (see pages 89–91 and 92–94). Bark can be quite brittle so check it by trying to break it with your fingers before using it.

Nuts
Chestnuts and acorns epitomize woodland gems. They are so perfect in shape and size and have a lovely smooth finish for satisfying, tactile jewellery. Once drilled there are countless ways in which to use them.

Seed cases
Acorn cups, spiky chestnut cases and other seed pods can be strung, twisted with wire and attached using headpins (see pages 74–75) to bring a quirky texture to your wild jewellery.

Leaves, feathers and bones

There is no end to a woodland's seasonal bounty, from the tiny flowers and wild garlic in spring to winter evergreens with their cheery, bright-coloured berries. Lush mosses and trailing ivies cling to trees all year round, while the rich colours of the autumn months provide plenty of riches that are all the more exciting because of their transience. A whole day spent in the woods will reveal a more unusual bounty provided by the many woodland creatures that one so often hears, but rarely sees – an elegant feather, perhaps, or delicate fragments of bone. With their many creative uses, these offer a welcome reminder of the birds and animals for whom the woodland is home.

Bone fragments
Bones can often be found in a woodland setting. Although stripped by other creatures, they need cleaning before you can use them for jewellery. This is very easy (see page 67) and well worth the effort as bone is a striking material to work with.

Fresh leaves
There are all sorts of different types of leaf to be found. They are ideal for making impressions in clay (see pages 95–97) or setting into epoxy resin: it really allows you to appreciate the lovely, intricate make-up of these natural treasures.

Dried leaves
As well as being used for their intricate patterns, autumn leaves provide a vivid shot of bright natural colour. They are great for stitching together to make a whimsical cuff, or for weaving into a twig base using silver wire.

Feathers
All sorts of feathers in various shapes and sizes can be found nestling on the woodland floor. You need patience to gather a large number, so they are best used sparingly, either as the focal element of a piece or as soft background for other components.

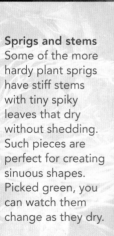

Sprigs and stems
Some of the more hardy plant sprigs have stiff stems with tiny spiky leaves that dry without shedding. Such pieces are perfect for creating sinuous shapes. Picked green, you can watch them change as they dry.

Moss and lichen
Moss and lichen have fantastic textures and colours. Soft and pliable when harvested, they are great for creating texture in wirework jewellery. They tend to go brittle when dry.

Storage and preparation

Some things you collect from the woods are best used straight away: you don't want them to dry out too much, as they become less supple and are liable to crack and crumble. This is true of the twigs for the free-form wirework technique on pages 76–77, which were cut and stored in water until ready for use. The same applies to moss, leaves and pine needles. Other materials, such as seeds, nuts, pods and cases are fine to collect and save for use later on: none of them will last forever, but you can allow these to dry out before you use them in your jewellery. Pieces of wood can be kept for ages: because you sand and treat them as you work them into your jewellery, you preserve the wood in such a way that it can last for a very long time, often improving with age.

Basic preparation

Most woodland materials are easily prepared for working into a jewellery design. Seeds, leaves and bark may simply need a quick wipe to remove any loose dirt before being trimmed. And, in the case of seeds and bark, you may need to drill them prior to use. Most feathers can be used after a brief wash in soapy water, while fragments of bone require more thorough cleaning. Preparing pieces of wood is also more involved, depending on how you intend to use them (see Working with wood on pages 68–69). However, apart from a multi-tool with various attachments and a junior hacksaw, there is very little need for any specialist equipment.

Preparing bone

Boil bone fragments in plenty of water to remove any vestiges of leftover meat. Even clean-looking bones may have been on the woodland floor for a while, so you would be advised to scrub them with wire wool and soapy water. Then leave them to dry before use.

Washing feathers

Found feathers can be made clean and hygienic by soaking them in warm soapy water. You could use a toothbrush to scrub the quill part of the feather gently. Lay feathers on some clean paper to dry out slowly and thoroughly before use.

Drilling seeds

1 With a little practice, it is easy to drill through the larger seeds and nuts, such as conkers, simply by holding the multi-tool in one hand and steadying the seed with the other. Use a slow speed and a sharp bit to avoid the drill slipping on the curved surface.

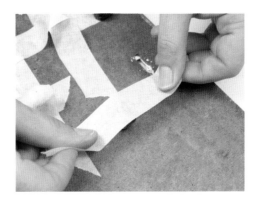

2 Alternatively, you may prefer to tape the seed or nut to a protected work surface using masking tape: use two strips of tape per seed. This secures the seed, allowing both hands free to use the drill. This would also be the method to follow if using a hand drill.

3 It is a good idea to mark the centre of the seed or nut with pencil before drilling, feeling the seed through the tape with your fingers. Taping the seeds in this way also speeds up the process in a project that requires a large number of them.

Working with wood

There are numerous ways in which to incorporate wood into your jewellery designs. In most cases, the wood is easily prepared using a handful of woodworking techniques. To some extent, the success of a piece will rely on the wood that you choose and whether it has an interesting grain or figuring. The main thing is to avoid rotten wood. This may sound obvious, but sometimes a piece that has been lying around on a woodland floor for a while will look lovely, but may crumble when you try to work with it. A good test is to see if you can snap the piece with your fingers.

wild jewellery

Working by hand

The drilling and sanding techniques used in this section can be carried out using a multi-tool with the correct attachments, but you can also drill and sand by hand. While a multi-tool sander is ideal for rough or speedy work, sanding by hand will give a smoother finish when it comes to completing a piece and exposing the natural grain. When sanding by hand, you simply work your way from coarse to fine sandpapers. Hand drills use the same size bits as a multi-tool. Should you be working on pieces that are especially delicate, a hand drill is also recommended, as it gives you greater control over speed and care.

Drilling

Use a multi-tool and an appropriate size drill bit. It is best to opt for quite a slow setting, particularly if you're trying to drill hardwood, so that the drill doesn't skid across the surface. Only once the drill has bitten, should you increase the speed to get through the wood.

Sanding

Sanding attachments for a multi-tool are good for use on jewellery projects, particularly for getting the initial shaping of a piece right. Some attachments are quite small and allow you to get into difficult spaces. They also have the benefit of being quick.

Sawing

1 If you have in mind how you want to use the wood when you're foraging, you can consider whether you want to cut across the grain or with the grain. If you cut with the grain, the pattern will be like long strips, while against the grain will produce shorter swirls or spots.

2 Wood knots render particularly attractive slices of wood, often with lovely swirly markings. A natural point of weakness in a branch, sawing a knot must be carried out carefully to avoid putting it under too much strain. Use a fine jeweller's saw to start with.

3 Work around the circumference of the knot, taking your time and sawing rhythmically to avoid the saw snagging. Once you have sawn around the circumference, change to a junior hacksaw. This will allow you to saw cleanly through the centre of the knot.

Threading techniques

As well as the methods described elsewhere in this book (see pages 22–23 and 24–26), there are a handful of threading techniques that work well with various woodland finds. Favourites include interweaving large nuts or seeds with a length of ribbon; making simple wire links to use smaller seeds as beads; and incorporating pine cones into a twisted-wire design. You can experiment with these techniques, introducing additional elements, to make striking necklaces, bracelets and earrings.

TOOLS AND MATERIALS:

- conkers
- ribbon
- tigertail
- crimp beads
- multi-tool and drill bit
- sewing needle
- flat-nosed pliers
- wire cutters

QUICK SIZE GUIDE

For a necklace, the tigertail should be the desired length of the finished piece, say 40cm. The ribbon needs to be longer to allow for the weaving in and out of the conkers – 50cm – and 2cm wide for a nice chunky look.

Interweaving ribbon

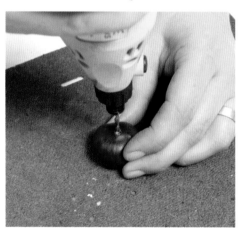

1 Twelve conkers were used for this project. Drill a hole right through each one, making sure your work surface is well protected. Tape the conkers down onto some old board if you're worried about them slipping (see page 67).

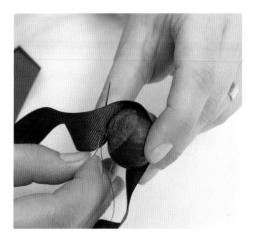

2 Thread the tigertail through the needle. Decide where the first conker should sit and push the needle through the ribbon at this point. In this example, this was 15cm from one end.

3 Pull the tigertail through the ribbon until you have about 2.5cm left at the end. Thread a crimp bead onto this short length and squeeze it closed using flat-nosed pliers.

4 Push the needle through your first conker to thread it onto the tigertail. Wrap the ribbon snugly around the bottom of the conker and take the needle through the ribbon to secure it.

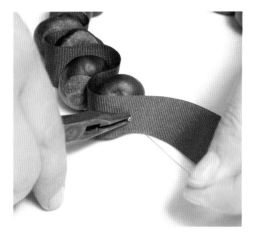

5 Thread a second conker on straight away, making sure the ribbon sits flat between this one and the first. Wrap the ribbon over the top of this conker and take the needle through the ribbon, before adding the next conker.

6 Continue adding conkers and weaving the ribbon in the same way. Once finished, simply thread another crimp bead onto the tigertail and squeeze closed to hold the conkers tightly in place. Trim any excess.

FINISHING TIPS

This technique is most effective carried out with the bigger seeds you are likely to find – conkers, chestnuts and acorns – so is most suitable for making bold statement necklaces. One of the best things about it is how quickly you can make a striking piece of jewellery without any need for a complicated fastening. If you use a long enough strip of ribbon, you simply tie the ends at the back of the neck with a luxurious bow.

into the woods

Making wire links

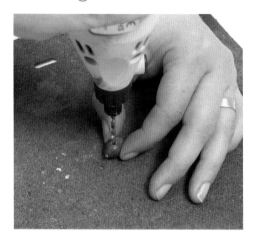

TOOLS AND MATERIALS:
- acorns
- 6-mm gauge silver wire
- multi-tool with drill bit
- flat-nosed pliers

QUICK SIZE GUIDE

These links work by making a small loop at each end of a piece of wire. The length of the piece of wire depends on the size of the seeds you are using. Acorns and similar size seeds need a 10-cm piece for each link. Increase this to 15cm for larger seeds.

1 Carefully drill a hole through each of your acorns (see page 67). Ten were used here, to make a simple bracelet. It is easier to drill the acorns horizontally rather than from end to end.

2 Thread a length of wire through the first acorn. Make a loop around 2cm up from one end and use flat-nosed pliers to wrap the short end around the base of the loop to close it.

3 Now make a loop around 2cm up from the opposite end of wire and use flat-nosed pliers to wrap the short end around the base of the loop in the same way. This is your first link.

4 Use a second piece of wire to repeat step 2 using the next acorn. Before repeating step 3, thread the loose end of the second wire through one of the loops you made with the first acorn.

5 Continue adding acorns until you have the desired length of linked seeds. Attach a simple hook clasp (see page 53) at one end of the chain, for linking through the last loop at the other end.

Twisted wirework

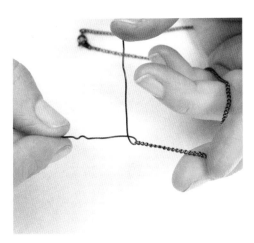

TOOLS AND MATERIALS:
- fir cones
- enamelled wire
- ready-made chain with fastening
- wire cutters
- flat-nosed pliers

QUICK SIZE GUIDE

For a focal piece on a necklace, use 50cm of enamelled wire. You could make a simplified version for a bracelet using 25cm of enamelled wire, and a ready-made bracelet chain.

1 Use wire cutters to cut the chain at the midway point. Thread one end of the enamelled wire through the last link of the chain and pull it through to the mid-way point of the wire.

2 Fold the enamelled wire in half to make a loop at the chain end and twist the two lengths of wire together for around 2cm using flat-nosed pliers. Separate the two lengths of wire again.

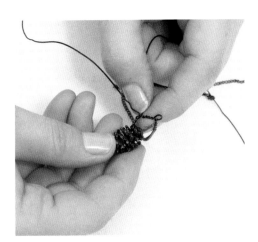

3 Take a small woodland find – in this case a fir cone – and wrap one length of wire around it, with 2cm of wire either side. Now twist two 2-cm lengths of wire together to secure.

4 Repeat along each length of wire until you have around 6cm of wire left. Loop one end through the last link on the ready-made chain, bring it back on itself and twist to complete.

5 This is a great technique for showcasing and suspending the more sturdy woodland finds. This piece shows a handful of little fir cones, with the stems kept on some of them.

Making and using headpins

TOOLS AND MATERIALS:
- selection of seeds or nuts
- 0.8-mm gauge silver wire
- wire cutters
- goggles
- small torch
- fireproof brick
- flat-nosed pliers
- plastic tweezers
- pickle
- sharp sturdy needle
- hammer
- metal plate

QUICK SIZE GUIDE

The length of wire you need for each headpin depends on the size of your seed or nut. Measure the depth of the piece you are threading and add an extra 3cm for attaching it to another component. The number of lengths you need will depend on the jewellery you are making.

Threaded through a seed or nut, headpins are used for attaching woodland finds to other jewellery components, such as necklace chains and ring shanks. The technique eliminates the need for any soldering, which might damage or burn your delicate woodland finds. Attached pieces have a freedom of movement that allows them to rattle or shake depending on how they have been used.

1 Select your seeds or nuts. The steps here use little acorn cups. Decide on how many you want to use and whether you want them all to be the same size or not. Cut a number of lengths of silver wire, according to your design.

2 Wearing goggles, light the torch and hold so that the flames rises vertically. Using pliers, hold one end of a wire just above the top of the inner blue flame. In a few seconds the end of the wire will glow and melt upwards into a small ball.

3 Cool in water and then leave in pickle for a short while. Follow steps 2 and 3 to make as many headpins as you need. In the meantime, prepare your seeds. Here, the acorns are removed from their cups and the stems trimmed.

4 Make a hole through the base of each piece, ready for threading. Acorn cups are soft enough that you can do this using a needle. Although true for many seeds, you may sometimes find that you need to drill a hole (see page 67).

5 Take a silver headpin out of the pickle, rinse it in clean water and dry. If you like, you can hammer the ball flat on a metal plate. Thread the headpin through the cup or seed so that the ball is on the inside or at the bottom end.

FINISHING TIP

6 Holding the ball end of the wire firm, use pliers to make a small kink in the opposite end of the wire. Thread the wire onto your piece of jewellery – in this case a rattle ring shank (see pages 78–79) – and then bend it back on itself.

7 The end of the wire should meet the kink you made in step 6 to close the loop. Do not close the loop too tightly, as you want the cup or seed to be able to move about a bit. Add more cups or seeds in the same way.

For an elegant necklace you could use a headpin to attach a single acorn cup to a large silver hoop, before threading the hoop onto a length of fine silver chain.

Free-form wirework

This is a really useful, free-form wirework technique that is great for 'controlling' found materials, allowing you to make natural things into pretty jewellery you can wear. You can use the technique for making rings and cuffs or, on a slightly larger scale, for focus bib necklaces and fairy-like tiaras.

TOOLS AND MATERIALS:
- handful of green twigs
- 0.6-mm gauge silver wire
- selection of small beads
- wire cutters
- flat-nosed pliers

QUICK SIZE GUIDE

Use a single twig – say 10cm for a ring, wrapping it around the finger a couple times. For a cuff, 20cm is a good length, while a bib necklace or tiara would need 30–40cm-length twigs. Work with 100–150cm lengths of wire.

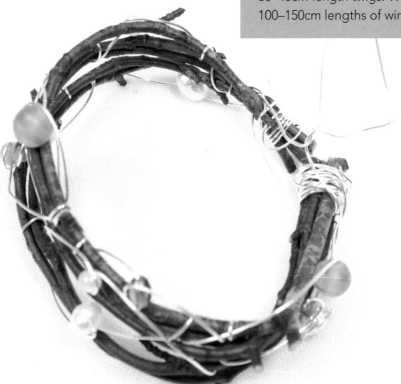

1 Select the number of twigs you'd like to use for your design – the cuff demonstrated on these pages has five. Using your wire cutters, cut off any leaves from the twigs and trim them down to an appropriate length.

2 Take a length of wire, measure about 50cm from one end and use this to bind the twigs tightly, 2cm up from one end. Use pliers to make a small loop in the wire and secure the base of this as you finish wrapping.

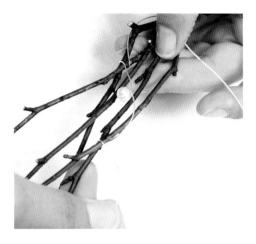

3 Loosely weave the twigs in and out of each other. Smooth the end of your wire into a curve with your thumb and finger and thread a bead onto the wire. Hold it in place by doing a half-wrap around the bead (from one hole to the other).

4 Now ease the wire into a curve in the opposite direction and wrap it around one of the twigs. Continue the curve and attach another bead in the same way. Wrap the wire around a different twig, keeping it fairly loose.

5 Continue working along the twigs in the same way until you reach the other end. Repeat step 2 to bind this end, making a second loop (and thus an easy means of fastening – say by attaching a length of ribbon to each loop).

6 Once both ends of the piece are bound, trim any excess lengths of twig. Work your way back along the piece in the opposite direction, attaching beads and weaving in and out of the twigs as before. Do not trim the wire when done.

7 Try the piece for size. Because the twigs are green and bendy, you can manipulate them to form the desired shape for your finished piece. Here they are eased into a nice oval shape to fit comfortably around the wrist.

8 Use the excess wire to hold the piece in shape while it dries. This may take 24 to 48 hours. Snip off any excess wire and work the loose end into the piece. It is then ready for attaching a length of ribbon or similar for tying when worn.

Making a 'rattle' ring shank

There is something of a free-form Celtic look to this thick, silver-wire ring shank, which means that it sits really well with all manner of woodland finds. If you make your own headpins (see pages 74–75) you can use them to attach little acorn cups, stacks of sycamore wings or little fir cones. Small treasures attached in little groups work really well, as they have the freedom of movement to rattle about a bit on the ring shank.

TOOLS AND MATERIALS:

- 1.5-mm or 2-mm gauge silver wire
- wire cutters
- mandrel (optional)
- flat-nosed pliers
- fireproof brick
- silver solder
- paintbrush
- borax cone
- goggles
- soldering torch
- plastic tweezers
- pickle
- metal plate
- hammer

QUICK SIZE GUIDE

The length of wire you need for the ring shank is enough to go around your finger, plus an extra 1cm at each end. This gives you enough to make the loop for attaching rattle pieces. The simplest way to measure your finger is by using a piece of string or strip of paper, which you then run along a ruler.

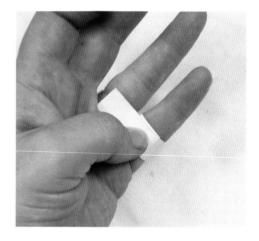

1 Use a strip of paper to calculate the length of wire you need, add 1cm at each end and cut it with wire cutters.

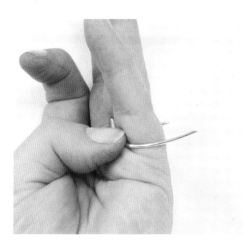

2 Bend the length of wire around the finger that the ring will be worn on or a ring mandrel to manipulate its basic shape. Centre the finger or mandrel along the length of wire, to allow for the excess at each end.

3 Using flat-nosed pliers, bend one end of the wire to form a loop big enough to thread the other end of wire though. Now bend the other end of the wire to make a loop in the same way, and so that both wire ends meet.

4 Wearing goggles, place the ring shank on a fireproof brick and solder each loop closed, following steps 3 to 6 on page 45. Drop the shank into some pickle and leave it there for a few minutes to remove any firestain.

5 Take the ring base out of the pickle, rinse it under water and dry thoroughly. Hammer the loops on the shank on a metal plate for a satin finish, if desired.

FINISHING TIPS

Small woodland finds, including these little fir cones, are great for attaching to the rattle ring shank. Consider the number of pieces you attach carefully, as you don't want the piece to be so crowded that there is no room for movement. These cones have been attached using the Twisted-wire threading technique shown on page 73. The headpins made on pages 74–75 are better suited to other woodland finds, such as acorn cups.

Making a 'tickle' ring shank

The tickle ring shank shows you how to make a quirky little ring that's sure to be a talking point wherever you go. Based on a simple band ring, the technique adds a short length of silver tube to the design, perfect for holding all manner of ticklish woodland finds. Part of the appeal of this piece is that you can change the tickler almost daily for anything that takes your fancy – a downy feather, sprigs of moss, even fern fronds.

TOOLS AND MATERIALS:

- 1-mm thick silver sheet
- 5-mm diameter tube
- paper
- pencil
- snips
- fireproof brick
- goggles
- soldering torch
- mandrel
- flat-nosed pliers
- silver solder
- paintbrush
- borax cone
- plastic tweezers
- pickle
- needle file
- wire wool
- metal plate
- hammer

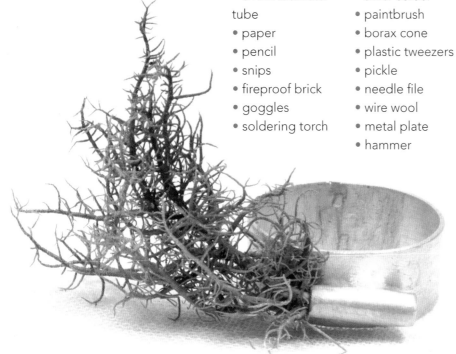

QUICK SIZE GUIDE

The length of silver sheet to cut is the same as the circumference of your finger – around 8cm. Proportionately, the design looks best if the silver tube is the same length as the ring is wide – a 1-cm bandwidth is solid-looking while still being comfortable.

Making the band

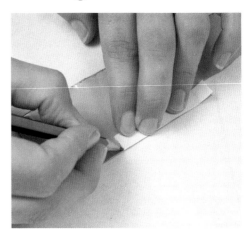

1 Measure your finger using a piece of paper and transfer this measurement to the silver sheet. Use snips to cut the silver to the same length and 1cm wide. If you're making this for somebody else, you'll need to know their ring size.

2 Place the silver on a fireproof brick. Light the torch and heat the silver until it glows red. Wear goggles for this and pass the flame to and fro over the silver to heat the strip evenly.

3 Allow the silver to cool down enough to handle comfortably, then bend the strip around a mandrel until the short edges meet. You may need flat-nosed pliers to help manipulate the silver.

4 Use flat-nosed pliers to push one edge of the ring past the other and allow them to spring back together. This will help to make sure that the join fits snugly for soldering.

5 Prepare the band for soldering. First cut the solder, then paint the join with borax. Place three pieces of solder along the join in the ring. (See steps 3 to 5 on page 45.)

6 Wearing goggles, light the torch, and pass the flame over the ring using a circular motion so that the metal heats up evenly. The solder should melt just after the silver glows red.

7 Cool the ring down and pickle. After pickling, rinse the ring in water and dry thoroughly. File the join smooth, then rub the ring all over with wire wool to give it a textured finish.

Adding the tube

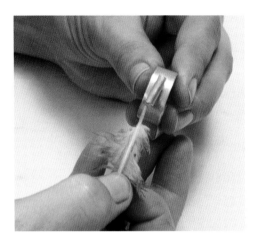

1 Flatten the silver band a little by tapping it lightly with a hammer. This makes it easier to place the silver tube. Paint some borax on the surface and put a few pieces of solder on top.

2 Balance the piece of tubing on top of the solder. Here, it sits horizontally, but could also be vertical. Solder, cool and pickle the ring. Sand and finish using wire wool for a textured look.

3 Ease your woodland find into the tube – in this case a feather. Place the ring on a metal plate and tap the top of the tube to flatten it slightly if you want to secure the feather permanently.

FINISHING TIPS

This is one of those projects that you can change regularly to suit the season or a favourite outfit. Perfect for incorporating a flower stem or plant frond, it makes the most of the transient nature of many woodland finds. Select a piece carefully – a sprig of wild lavender or a piece of soft moss and you can watch your jewellery change as it transforms into its dried state. Make sure you keep stems nice and long to maximize the tickly effect of the design.

Making a lariat-style chain

Ring the changes with this simple chain design in which the fastening becomes the focal point of a necklace. Like the Tickle ring shank on pages 80–82, it is one of several techniques in this section that is well suited to showcasing the transient nature of many woodland treasures. As long as you don't hammer your woodland finds in too hard, you can ease them out and exchange them at whim.

TOOLS AND MATERIALS:
- 1.5-mm gauge silver wire
- 0.8-mm gauge silver wire
- fine silver chain
- 5mm silver tubing
- wire cutters
- mandrel
- flat-nosed pliers
- metal plate
- goggles
- soldering torch
- silver solder
- paintbrush
- borax cone
- plastic tweezers
- pickle
- hammer
- emery paper

QUICK SIZE GUIDE

In this design, the woodland find needs to hang vertically to keep the tension in this design, so the technique is best suited to making necklaces. A length of chain ranging from 40–50cm should work. The hoop is around 8cm, while a 1.5cm silver tube will hold most woodland finds.

Preparing the components

1 Make the silver hoop. This may be a perfect ring or an irregular pebble shape. If you want a true circle, wrap the 1.5-mm gauge wire around a mandrel. Otherwise, use a pair of pliers to manipulate the wire by hand.

CONTINUED

2 Place the hoop on a metal plate. Wearing goggles, light the torch and solder the hoop closed, following steps 3 to 7 on page 45. Cool the ring under water, holding it in your pliers, before dropping it into some pickle.

3 Using plastic tweezers, remove the large silver hoop from the pickle and wash and dry it thoroughly. Place it on a metal plate and hammer it flat to give it a textured finish. If you prefer a smooth finish, you can leave this step out.

A CLEVER DESIGN

The lariat-style chain is a simple, yet clever, technique. The success of the resulting necklace depends on the ability of the focal element to counterbalance the weight of the chain once the piece is assembled. This example uses a feather, but you can use anything as long as it is heavy enough to hang without being pulled back through the hoop – a small bunch of pine needles, a leaf, even a piece of nicely textured bark.

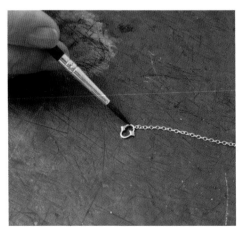

4 To prepare the silver tube element, start by making a small ring or oval using around 1.5cm of 0.8-mm gauge silver wire. Keeping it open at this stage, thread the ring through one end of the silver chain.

5 Cut two small pieces of solder and place one of them over the join in the small silver ring. Place the second piece opposite and balance the piece of silver tubing on top so it will fuse to the ring when you heat it with the torch.

6 Carefully heat the piece until the solder melts. Keep the flame moving and work slowly to prevent the silver tubing from falling off the solder. Pick the piece up using pliers, cool under water and place in the pickle.

wild jewellery

Assembling the chain

1 Repeat step 4 on page 84 to make a second 0.8-mm gauge silver ring or oval, and use this to attach the silver hoop to the other end of the fine chain. You may find this easier to do using pliers to thread it through the chain.

2 Return the silver hoop to the metal plate and solder the small silver link closed. Cool it off in water and put it back in the pickle for a few minutes to clean off the fire stain. Rinse under clean water and dry.

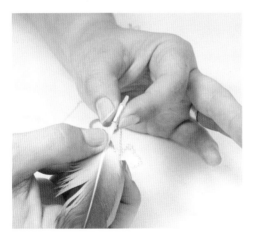

3 Prepare your woodland find (see pages 64–65). Whatever you use, make sure the decorative piece has a good length of stem, which you should push gently into the bottom of the tube.

4 If it does not fit snugly, simply tap the tube a couple of times with a hammer to flatten it a little. Go over the tube and links with emery paper for a nice, smooth satin finish.

FINISHING TIP

This is a technique that suits a woodland find with some length. Because it needs to hang vertically, it really adds to the success of the design if the natural element is also long and slender, as in the example below. It must also be a piece that has a long enough stem to insert into the silver tube. Look out for pieces that have a chunky enough stem to wedge into the silver tube without the need to hammer it, as this will make it easier to swap an old find for a new one every so often.

Making resin beads

Slices of twig make great beads. Add interest by making circular indentations in the face of each slice and filling with epoxy resin. You can colour the resin, using a single colour per piece or combining different shades. Larger slices of twig can be resin-filled in the same way to make a striking pendant for a necklace or the centrepiece for a brooch.

TOOLS AND MATERIALS:
- chunky twig
- two-part epoxy resin
- old powder-based make-up
- olive oil
- wood saw
- multi-tool with round burrs, sanding attachment and drill bit
- cocktail stick
- soft cloth

QUICK SIZE GUIDE

For reasonably chunky threading beads, find a twig with an approximate diameter of 2cm. A length of 10cm will make ten 1-cm beads. Make sure you find a length of twig for the number of beads you wish to make. A branch slice with a diameter of 5–8cm would make a good-sized pendant or, perhaps, a centrepiece for a brooch.

1 Using the saw, cut a number of slices from the twig – as many as you need depending on the design you might have in mind. This example is for a necklace that uses seven to eight twig beads strung with tigertail.

wild jewellery

MIXING EPOXY RESIN

Squeeze out equal amounts of resin and hardener beside each other on a non-porous surface. Mix together quickly, using a cocktail stick.

Add a colour of your choosing. Use things you might have lying about the house: old make-up, shoe polish, watercolour paint and so on.

Still using the cocktail stick, mix the colouring into the epoxy resin. Work quickly as the resin will begin to set once it has been mixed.

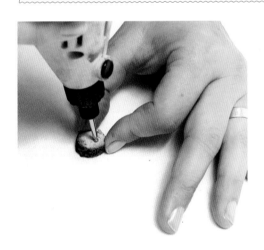

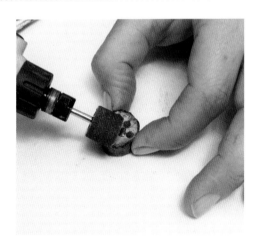

2 Using the round burr on the multi-tool, make random indentations in one cut face of each twig slice. You can use a Phillips screwdriver if you don't have a multi-tool. Two different sizes have been used here for greater variety.

3 Mix your resin (see above) and use the cocktail stick to blob a small amount of resin into each cavity in the beads. You needn't be too neat, as you'll sand the surface later on. Leave to set really hard in a warm, dry place for 48 hours.

4 Use the multi-tool with sanding attachment to sand the resin down. You could do this by hand with sandpaper. If you use the electric tool, don't work on it too long as the resin can heat up and start to go soft again.

CONTINUED

5 Sand the other side of each twig slice so that the backs of the beads are nice and smooth (this is quite important as this surface will rest against your skin).

6 Drill a hole through each twig slice from top to bottom (see page 69). This will allow them to sit flat against the skin when threaded together.

7 When you've drilled all the slices, it's a good idea to oil the wood to protect it and bring out the colours a bit. Use olive oil on a soft cloth and rub it gently over the surface of each bead.

8 Use any threading technique to incorporate the beads into a striking piece. One of the most secure methods is by using tigertail and crimp beads following the steps on pages 24–25.

FINISHING TIPS

You can use resin beads in various ways. In the steps shown, the beads are threaded randomly alongside dark, stone beads for a solid-looking necklace. Singly, the beads can become a unique focal point for a brooch or ring. For example, make a brooch by placing a twig bead onto a beautifully textured piece of found bark. You can do this by drilling two holes through the bark and pushing the wires of a threaded bead through from front to back. Twisted together to secure the bead in place, the wires can then be wrapped around a small brooch back to finish. Another pretty use for a single twig bead, is by threading it onto a silver wire ring shank (see pages 30–31), as shown below, with iridescent beads of a complementary colour.

Making a silver and resin pendant

This technique demonstrates a simple method for making a striking focal piece using epoxy resin. Fundamental to the technique is that the device for attaching the pendant to a necklace chain – in essence a rectangle of sheet silver – is set within the resin to make an all-in-one component, ready for threading. Working with bone for this example, the technique is equally successful using wood. You could also make a brooch or focal piece for a bracelet, altering the design of the silver element to suit.

TOOLS AND MATERIALS:

- section of bone, cleaned and dried (see page 67)
- 1-mm thick sheet silver
- 1.2-mm gauge silver wire
- epoxy resin
- old make-up
- small hacksaw
- face mask
- goggles
- snips
- metal plate
- hammer
- needle file
- multi-tool with 1-mm drill bit and sander
- small clamp
- flat-nosed pliers
- fireproof brick
- silver solder
- paintbrush
- borax cone
- goggles
- soldering torch
- plastic tweezers
- pickle
- masking tape
- cocktail stick
- sandpaper
- emery paper

QUICK SIZE GUIDE

The size of the slice of bone you need depends on the piece of jewellery that you are making. A diameter of 3–5cm makes a good-sized pendant or brooch. A focal piece for a bracelet or cuff might be 1–2cm in diameter. The depth of a piece is up to you. The technique works well for the more chunky designs, so a minimum of 1cm is good for all pieces of jewellery.

Preparing the bone

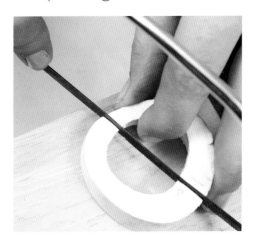

1 Using a small hacksaw, cut a slice of hollow bone. Wear a face mask and goggles, as bone dust can be harmful. Cut a groove right across the bone slice with your saw, ready to put the silver element into.

Making the silver element

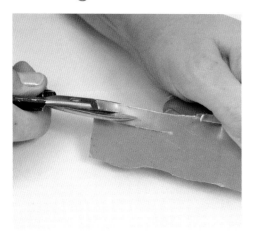

1 Using snips, cut a piece of sheet silver that is about 0.5cm wide. The length of the silver needs to match the diameter of the bone slice.

2 Place the silver on a metal plate and hammer it gently all over to flatten it out. You can file any really rough edges using a needle file.

3 Use the multi-tool to drill a small hole in the silver, about 3mm in from one end. Use a clamp to prevent the silver from slipping about as you work.

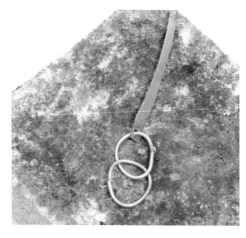

4 Cut two 5-cm lengths of 1.2-mm gauge wire and form each into a silver ring. Use flat-nosed pliers to help manipulate the wire.

5 Link the two rings and thread one through the hole in the sheet silver. Solder the joins on the rings following steps 3 to 7 on page 45.

6 If you prefer a textured, matt finish, you can hammer the rings flat. Place them on a metal plate to protect your work surface.

Filling with resin

1 The silver element is now ready for positioning in the cut groove in the slice of bone. It should fit snugly, so hammer it in gently if necessary.

2 Using masking tape, tape across the top face of the bone slice, overlapping the strips carefully, and place tape-side down onto a non-flammable surface.

3 Mix up your epoxy resin (see page 87) and mix your chosen colour into the resin using a cocktail stick. It does set quite rapidly, so act quickly.

4 Pour the resin into the bone slice so it falls equally on both sides of the silver element: it can create heat so make sure your work surface is protected. Leave to set for at least 48 hours.

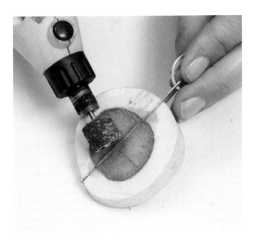

5 Use the sander on your multi-tool, followed by emery paper, to polish the resin for a smooth finish. You can also use a coarse sandpaper followed by emery paper, if working by hand.

FINISHING TIPS

Thread your finished piece onto ribbon or chain to wear as a long, retro-looking pendant – a length of 50cm should do it.

Inlaying silver

This is a straightforward technique for inlaying silver wire and tube into a piece of found wood. These steps lead to a beautiful pendant, but you could use the technique for any focal element of your work – the centrepiece in a tiara, a striking brooch or a pair of matching earring drops. Use cold connections, like those on pages 52–53, to join the wood to perspex or handmade silver chain (see pages 48–51).

TOOLS AND MATERIALS:
- piece of found wood, sanded smooth (see page 69)
- 2-mm gauge silver wire
- 5-mm silver tubing
- epoxy resin
- multi-tool with different size drill bits/grinding/cutting and sanding attachments
- pencil
- wire cutters
- small hacksaw
- wooden bench peg
- metal plate
- hammer
- olive oil
- soft cloth

QUICK SIZE GUIDE

The quantity of silver wire or tube you need depends on the design of your piece. First work on the design, deciding on how many places you want to inlay with wire and/or tube. Multiply the number of holes by the depth of the wood to get a rough idea of the length you need and allow a little extra to be on the safe side.

Preparing the components

1 Look for a branch that will give an interesting cross section when cut. Perhaps you have a piece with a knot or burl that you can saw through, as here.

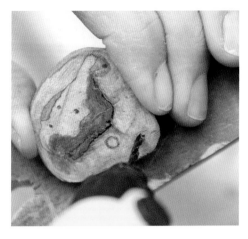

2 Saw off a chunk or slice of wood to the size you need (see page 69) and sand it smooth. Depending on the piece, you can do this by hand (see page 69), but you may find it quicker and easier to use a multi-tool with a sandpaper attachment.

3 Place the piece of wood on your work surface and use a pencil to mark holes for inlaying silver wire and silver tube, according to your design. If you intend to attach the piece to a length of chain, or cold connections (see pages 52–53), mark holes for this, too.

4 Using the multi-tool with a 2-mm drill bit, or a hand drill, carefully drill through the holes for the silver wire and top holes for the chain hanging (see page 69). Use a 5-mm drill bit to make the larger holes for the silver tube. Protect your work surface throughout.

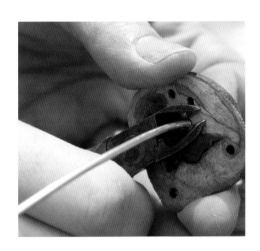

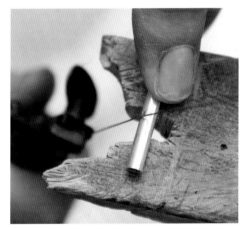

5 Starting with the silver wire, push it into one of the holes so that the end is flush with the back of the piece of wood. Now snip it flush with the surface on the front. Repeat this for all of the 2-mm holes and set the silver pieces aside.

6 Rest the piece of wood on your work surface and hold the length of silver tubing next to it. Use a pencil to mark the depth of the wood onto the piece of silver tubing. (You can use the first piece of tubing to measure any more.)

7 Carefully cut the tube using a hacksaw and following the pencil mark as your guide. This can be quite tricky, and is best carried out with the length of silver tubing laid across the width of a wooden peg for support.

Inlaying the silver

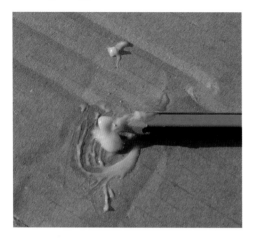

1 Mix up a small quantity of epoxy resin following the steps on page 89. In this example, the epoxy resin is used in its clear, uncoloured state, but you can add a colour to it if you like.

2 Use the end of a pencil to push a little of the epoxy resin into each of the holes intended for inlaying, and around the openings. This just helps to hold the silver inlay in place.

3 Push each length of wire and tube into its prepared hole and use a hammer to tap them into place in the wood. Use a gentle action and rest the piece on a metal plate to do this.

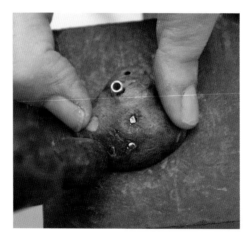

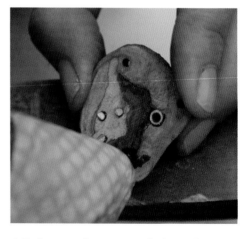

4 Depending on the piece of wood (this example is not of a uniform depth all over), the silver may protrude in places. Use a grinder tool to give them a smooth finish.

5 Turn the wood over so that the face is laid down on the metal plate and gently hammer the wire and tube. This rivets them a little, helping to secure them in the wood.

6 Dab a small amount of olive oil onto a clean cloth and rub it over the wood to bring out the grain. Your piece is now ready for incorporating into your final jewellery design.

wild jewellery

Working with Precious Metal Clay

Precious Metal Clay (PMC) is a clever medium, developed in Japan in the 1990s. Silver particles are mixed with a binder and water to make a malleable clay material. When fired, the binder burns off to reveal just the silver form. It can be used with all manner of textured woodland finds – leaves, bark, seed pods – to create decorative embellishments for your jewellery.

TOOLS AND MATERIALS:
- Precious Metal Clay (PMC)
- a few small leaves
- 10 playing cards
- small rolling pin or similar
- craft knife
- paintbrush
- plastic lid
- cocktail stick
- metal tray
- fireproof brick
- goggles
- small torch
- emery paper
- needle file

QUICK SIZE GUIDE

PMC is pricey, so you want to be sparing with it. Around 14g will produce four to five small leaves. If you plan to make little silver links for joining the leaves, allow for 1cm of 0.8-mm gauge silver wire per link.

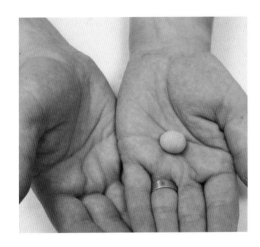

1 Work a small ball of PMC in your hands to soften it, as you would regular modelling clay. Make sure your hands are clean and roll the clay until it has warmed up sufficiently to use.

CONTINUED

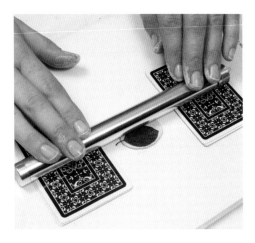

2 Place the PMC on your work surface with five playing cards stacked either side. Balance the rolling pin on the cards and gently roll out the PMC. Remove a card, from each stack, centre your leaf on the clay and gently roll it.

3 You will see that, by removing the playing cards, you are able to make a leaf-shaped imprint in the PMC, complete with veins. When you've finished rolling the clay, carefully peel the leaf off and discard.

4 Cut around the leaf shape with a craft knife. Gently lift the excess clay away and roll it into a ball, ready for making the next leaf (put excess clay back in the bag so it doesn't dry out).

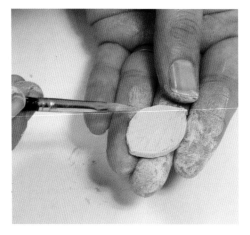

5 Very carefully smooth the edges of the leaf with your fingers. You can apply a few drops of water with a paintbrush, but avoid wetting the top, as you may lose your leaf imprint.

TIPS FOR USING PMC

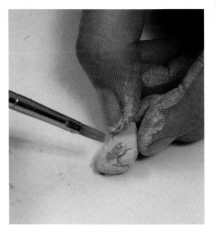

■ Precious Metal Clay is very quick drying and will become unusable if allowed to dry out, so prepare your work space in advance.

■ Always check the manufacturer's instructions on drying times, as this will give you a good idea as to how long you have to work with the clay. It is good practice to remove only as much as you need at a time, returning off-cuts to an airtight bag as soon as you can.

■ Re-use small off-cuts by rolling several into a single piece, and adding a little water to bind them.

■ The playing cards help to roll to a constant thickness – a minimum of 3mm is recommended, as the pieces become thinner when fired.

■ Oven times and temperatures vary, so check the instructions. These pieces went into a 150°C oven for 20 minutes.

wild jewellery

6 Repeat the same process to make 3 or 4 more leaves, placing them on a plastic lid as you complete them. You can take another playing card away from each stack if you want to achieve a deeper imprint.

7 Using a cocktail stick, make holes in the leaves. A bracelet charm will need a hole at each end, and a pendant just one at the top. Transfer the leaves carefully to a clean metal tray and place in a very low oven (see Tips, opposite).

8 Allow the leaves to cool, then put them on your fireproof brick ready to fire. Wearing goggles, heat each leaf one at a time, just until it glows red. Do not overdo the firing as you risk melting the silver and obliterating the imprint.

9 Firing the pieces will have burned off the clay binder. Cool the pieces in water, dry them and use fine emery paper to clean off any remaining whiteness. Finish each silver leaf by filing any rough edges with a file.

FINISHING TIPS

There are many ways to incorporate PMC pieces into your jewellery designs. One simple method is to make small silver links for attaching the pieces either to each other for a bracelet or necklace, to a chain as a charm or pendant or as drops for a pair of earrings. As mentioned on page 95, 1cm of fine-gauge silver wire is enough for each link. Manipulate the wire by hand to form a loop, and thread the PMC on before soldering the loop closed.

Woodland inspiration

More than any other environment that you are likely to explore for inspiration, a woodland setting is probably the most aesthetic. Completely surrounded by nature, all of the colours, tones and textures have a perfect synergy with one another. Choose pieces for their sculptural qualities, combining them with elaborate silver wirework, or juxtapose contrasting textures and colours in a piece to emphasize the beauty of the natural forms with which you are working.

Woodland stems
Silver-wire crochet makes a stark contrast with the rust-coloured needles. (See pages 34–37 for crocheting.)

Colour coordination
A necklace captures the subtle colours of a woodland setting. (See pages 112–113 for making paper beads.)

Wild choker
Silver mimics the untamed, tangled nature of soft, green moss. (See pages 76–77 for free-form wirework.)

Fairy-tale tiara
Elegant free-form wirework techniques have been used to create this delicate piece (see pages 76–77).

Spruce decoration
Pine needles worked into
a necklace using free-form
wirework (see pages 76–77).

Acorn earrings
A pair of hook earrings,
made by manipulating silver
headpins (see pages 74–75).

Frosted bead
Moss set in smooth resin
offsets gnarled wood for this
necklace (see pages 126–128).

Mossy pendant
A moss pendant with steely
beads. (See pages 126–127
for making resin pieces.)

Jewellery by Justin Duance

Justin Duance designs and creates stunning wood and inlaid silver jewellery. Operating from his workshop in the fishing village of Newlyn, in Cornwall, he uses locally sourced and tropical woods, which he combines with precious metals in a sleek, contemporary way. Contrasting with the inlaid metal, the unique figuring of the hardwood or fruit wood in a piece is displayed at its best. Among Justin's signature pieces are his chunky, banded rings and perfectly rounded 'pebble' pendants. Each piece is crafted by hand to achieve the lovely smooth finishes that characterize his work.

Justin's jewellery is not simply remarkable on account of its beauty, however. For him, the choice of wood is very important, and each different type is imbued with symbolism. Oak, for example, represents strength and apple stands for love. The effect of this, when combined with a precious metal, can be of great personal significance to the wearer.

Clockwise from above:
An oak 'pebble' inlaid around the perimeter with silver. (See pages 92–94 for inlaying techniques.)
A melted, textured sterling-silver ring with a single band of oak inlay.
Sterling-silver cufflinks with striking, banded kingwood inlay.

Urban day out

Urban environments provide rich pickings when it comes to sourcing materials for making wild jewellery. Not only are there city parks and gardens in which to seek out natural treasures – spring flowers, delicate blades of grass or autumn berries – but there are man-made materials to be collected for recycling. The innovative techniques in this section include fusing plastic bags to make colourful sheet plastic, setting urban finds in polymer clay and resin, and making beads from paper and cardboard. One fascinating aspect of using urban materials is that many of them are transformed through the jewellery-making process. They take on new shapes, colours and textures that make them things of beauty in their own right.

An urban forage

Living in a built-up environment doesn't mean that you can't find beautiful things with which to make your wild jewellery. You don't have to wait until the next time you go to the beach or rely on a day out in the woods for a good forage. If you're feeling creative, chances are you'll want to get to work immediately. Next time you take a walk through your city, go with an open mind and you may be surprised at the wealth of found materials that you take home.

If you spend a lot of your time in a built-up city centre with not much greenery, there are plenty of found things that you probably come across all the time, and that are brilliant for recycling into contemporary jewellery. Just think: plastic bags, ring pulls from drinks cans, lost nuts and bolts, discarded inner tubes, paper, cardboard and plastic bottles. You'll see them everywhere.

Buried treasure

There is also a rich, historical aspect to living in a city, with years behind you of previous city-dwellers. Dig about in your garden and you might unearth chunks of antique crockery, old clay pipes and even, if you're lucky, old jewellery! Plenty of these found materials can be incorporated into your designs using the techniques in this chapter, but also those in the beach chapter of this book (see pages 12–57). And while you are in a city park, there is no reason why you cannot forage for more typical woodland finds, such as pine cones, acorns and twigs (see pages 58–101).

Getting started

The best way to go about an urban forage is simply to be more aware of your surroundings as you go about your everyday life. Look more closely at the environment in which you live and take a different view of what you see. Check pavements for ring pulls or mangled bits of metal as you walk to the bus stop or train station; pick delicate bits of grass from the verge outside your workplace to encase in resin; or rescue a plastic bottle from the recycling bin to make into a beautifully sculptural, curled pendant.

Using found materials in unexpected ways really does add a unique, original facet to your wild jewellery, keeping it cool and contemporary. And there's nothing to stop you honing traditional skills using metal, wire and threading techniques to showcase these urban materials. Whichever way you go about it, you will always succeed in making jewellery that has a quirky twist and good green credentials. You'll create striking pieces from materials that most people simply consider to be rubbish.

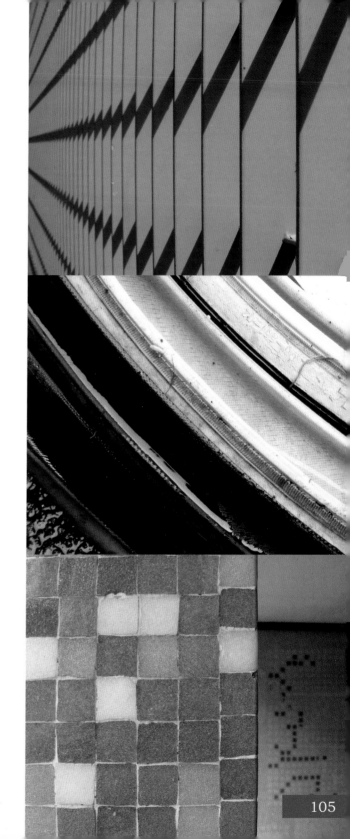

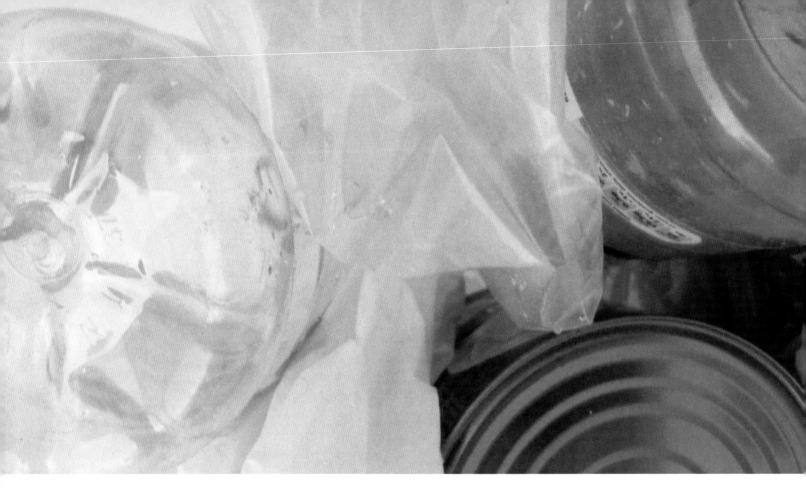

Metal, plastic and rubber

One man's waste is another man's treasure, and this has never been more true than on an urban forage. Start with a local recycling area and you'll be rewarded with plastic bags, bottles and tin cans in many bright colours. Scour the streets and it's a different story. Once you start looking there will be no end to the stray nuts and bolts you come across, bits of rubber from shredded tyres, fragments of broken reflectors, discarded bottle tops and ring pulls. With a little imagination, all of these can be put to much better use in your one-off, urban jewellery designs.

Plastic bottles

Plastics can take on a jewel-like appearance when incorporated into wild jewellery. You can make quite ethereal, organic shapes simply by cutting strips from a plastic bottle and heating them in a low oven. The edges curl into soft petal shapes.

Found metal

Smaller pieces of found metal – nuts, bolts, screws and pipe clips all make original embellishments, especially when misshapen and rusty. They look great combined with wire or resin, or when set into polymer clay (see pages 124–125).

Reflectors

Saw up found or recycled bike and car reflectors to create any shapes you like. They are easy to drill for threading (see page 111), or can be set into a claw setting (see pages 129 –131). Great fun for evening wear, they reflect light with plenty of sparkle!

Plastic bags

Most types of plastic bag can be fused together in several layers to create unpredictable, abstract designs and flat beads (see pages 116–117). They can also be crocheted to make cuffs and statement necklaces (see pages 118–120).

Old tyres

Worn tyres or inner tubes can be cut up to make striking, hard-wearing components for jewellery-making. You can rivet pieces together for tough-looking brooches, cuffs and necklaces (see page 134).

Drinks cans

Cut up drinks cans with sharp scissors and use the bright colours to provide lovely bracelet and pendant charms. Rusty cans work well for tougher-looking pieces. Sand the edges well before using.

Paper, card and park finds

A forage in a city park can yield some surprising results. Obvious treasures to seek out are tiny pretty flowers in spring and lush, polished berries in autumn. Leaves, twigs and seeds will also feature throughout the year and can be incorporated into designs using techniques in the woodland chapter. But you'll also find park pebbles and discarded newspapers and magazines. If the park has a café, there is every chance they will also have cardboard boxes for recycling. All of these finds are perfect for making beads or setting in resin to make striking focal elements for your wild jewellery designs.

Flowers

Little flowers and petals that you might notice on a forage in the local park can be preserved as a feature for striking jewellery using resin (see pages 126–128). They don't need to be big – the simpler shaped flowers work beautifully.

Paper

Glossy magazines and newspapers are ideal sources for making paper beads (see pages 112–113) and origami flowers. Think about colour and finish. You can finish pieces with clear wood varnish or PVA glue for greater durability.

Grass

Short, spiky blades of park or garden grass can be used to create more contemporary, linear designs in resin (see pages 126–128) or unusual textures in polymer clay. Seed heads from some grasses can look very pretty and delicate set in resin as well.

Park berries

Transient jewellery designs might include berries twisted with fine enamelled wire (see page 73), which will last for up to a month. For something a bit more durable, you could set them into resin to preserve them forever (see pages 126–128).

Park pebbles

Seemingly unattractive urban stones, broken glass or lumps of tarmac found can look striking set into a claw setting (see pages 129–131) or crocheted in wire (see pages 40–41). Small, flat pieces work best.

Cardboard

Thick card can be used to make chunky, concertina-type beads, simply by weaving two long, narrow strips together (see pages 114–115). It shows off the lovely crinkled corrugations.

Storage and preparation

Urban finds are not always the cleanest. Quite often they have been rolling about on the ground for some length of time and will be either dirty or rusty. For many materials – plastic bottles and bags, aluminium cans, reflector fragments, rubber – you can improve them simply by giving them a quick wash in some warm, soapy water. Nuts and bolts or pieces of sheet metal may need a little more attention, such as filing or sanding. In fact, it is very important to wear away any sharp edges on pieces of metal, as cuts and grazes caused by rusty metal can easily become infected. As with other wild finds (see pages 20 and 66), it is important to store urban finds somewhere they can be kept dry. Soggy papers and cardboard can go mouldy, and metals will rust further in a damp environment.

Sanding rusty metal

Washing plastic

1 The rust on a metal piece can be an important element in the design of a piece of jewellery, in which case, leave it as it is. Should you wish to remove rust, however, it is best to begin by using the sanding attachment on a multi-tool to soften the worst of the sharp edges.

2 With the sharpest edges removed, go over the whole surface by hand, using sandpaper or wire wool. You can achieve a smooth finish this way, although you might also take the shine off the surface. This can enhance the 'battered' nature of an urban find.

Washing plastics is very straightforward. Simply fill a bowl with warm, soapy water and clean off any dirt. You can use a sponge to wipe the surface, rather than submerge the whole piece if the soiling is light. If you have a number of very dirty pieces, use the kitchen sink.

Drilling plastic

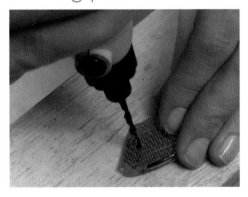

Sanding plastic

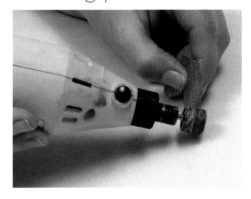

Using fused plastic

Drilling plastic must be done with care. Wear goggles in case a piece splits or cracks and little fragments go flying. Sharp drill bits make it easier to drill without this happening, while covering a piece with masking tape will help prevent the drill slipping (see page 67).

The sharp edges of plastic fragments can be sanded using a multi-tool. Also, it is not unusual for plastic to melt when drilling holes, leaving a rough or sharp surface when it solidifies again. It is a good idea to sand around the whole area to make it feel smooth again.

Fusing plastic bags together produces versatile plastic sheets (see pages 116–117). Use scissors to cut pieces from the plastic leaving a nice clean edge. For attaching to bracelets, earrings, necklaces and brooches using jump rings, simply pierce using a hole punch.

Making paper beads

Using paper to make pretty beads is one of the simplest jewellery-making techniques you can try your hand at. Think about a colour scheme for the piece you are making and find pictures in magazines that fit in with it. Depending on the paper you use, you can opt for a matt finish or a glossy one – it is up to you.

TOOLS AND MATERIALS:
- old magazines
- scissors
- crochet hook or pencil
- paintbrush
- PVA glue

QUICK SIZE GUIDE

The length of each bead will be the same as the base of the paper triangle you cut. The longer the strip, the fatter the bead. Experiment with different size triangles, depending on the piece you are making. For a necklace, aim to make 10 to 15 beads, and 5 to 8 for a bracelet, depending on length.

1 Cut a long strip of paper for each bead. Each needs to be slightly tapered at one end, like a tall triangle. Cut the number of strips you are likely to need. You can always cut more if you run out.

2 Wrap the base of a paper strip tightly around a crochet hook or pencil, just once. Paint a little PVA glue in front of this and continue to roll the paper strip, tightly, to the end.

3 As you approach the last ½cm or so of paper, paint on some more PVA glue so that you seal the paper bead with the last roll. Carefully remove the rolled paper from the crochet hook.

4 Paint glue over the entire surface. This seals the paper in position, but will also give the bead a harder finish once dry. Set the bead aside while you make more beads in the same way.

5 Once they are dry, you can snip your beads into shorter lengths using large, sharp scissors. You could also make smaller beads by starting with shorter, narrower strips of paper.

FINISHING TIPS

Alternate paper beads with found pine cones and coordinating glass beads for a pretty necklace. These are threaded onto tigertail.

A simple ring, made using the Wire ring shank method described on pages 30–31. The paper bead spans the width of the shank.

Here, paper beads are attached as earring drops. The pull-through chains can be made using the technique on pages 44–45.

Making cardboard beads

Beads made with cardboard are chunky and best suited to large necklaces. Paint the strips before assembling the beads to add some colour. The paint might crack a little, but this will add to the decorative effect. You could even seal it with varnish. You can use the same construction technique to make more delicate, colourful, paper beads for bracelets and earrings.

QUICK SIZE GUIDE

Good-sized beads can be made using 10–15cm lengths of card, each about 1cm wide. To thread the beads for a necklace, use around 50cm of tigertail.

TOOLS AND MATERIALS:

- corrugated cardboard
- ruler
- pencil
- craft knife
- PVA glue
- sticky tape
- small, sharp scissors
- large, sharp needle
- tigertail and crimp beads
- flat-nosed pliers

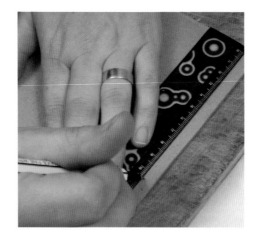

1 Decide on the size of beads you want to make. Here, the strips of card are 10 x 1cm. Trim your piece of card to the correct length using the craft knife.

2 Make sure that you cut the 1cm strips of card across the corrugations so that you can see the crinkled edge all along the long side of each strip.

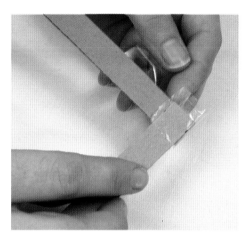

3 You need to use two strips of card for each bead that you want to make. Start by dabbing the end of one strip with a blob of PVA glue.

4 Lay a second strip over the first, with the glue end at 90 degrees. Leave the glue to dry or tape over the end to prevent the card slipping as you fold it.

5 Fold the first strip over the top of the second. Now bring the second strip back over the first, and so on, to create the concertina effect.

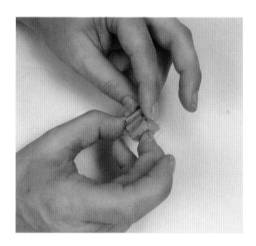

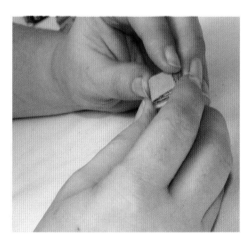

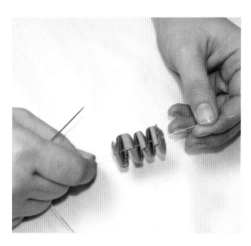

6 Continue until you reach the ends of the strips of card. Hold them in place with a blob of glue and secure with a piece of sticky tape while they dry.

7 Make as many concertina beads as you need, following steps 3–6. When they are all dry, you can remove the tape and trim them using scissors.

8 The best way to thread cardboard beads is to use a strong needle and tigertail. Follow step 2 on page 24 to make a loop at each end to secure.

Making fused plastic

This technique offers a very effective way of making sheet plastic from which to make all kinds of wild jewellery. Depending on the colours of the bags you use, you can produce anything from stylish black-and-white charms for a bracelet to the sugary, pastel colours of a layered brooch, and an all-out brightly coloured cuff.

TOOLS AND MATERIALS:
- plastic bags
- iron
- clean sheets of paper
- scissors

QUICK SIZE GUIDE

Before tearing any plastic, think about the scale of the jewellery you are making. Ideally, your pattern should be appropriate for your design. You might tear thumb-sized pieces for charms on a bracelet, for example, or palm-sized pieces for a large cuff.

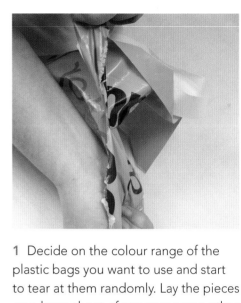

1 Decide on the colour range of the plastic bags you want to use and start to tear at them randomly. Lay the pieces on a large sheet of paper or some other surface that you can later iron on.

2 Arrange pieces as you work, moving them around to get a pattern that you like. You can have fun experimenting at this stage with a random, abstract design, or one with more structure.

3 When you are satisfied with your design, lay a second sheet of paper over the top of your plastic. Using a medium setting without steam, pass an iron slowly over the paper.

4 Avoid making contact with the plastic, as this will result in a sticky mess! Pass the iron over the paper a couple times, and peel back the paper while warm, to prevent the plastic sticking to it.

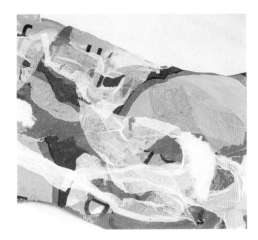

5 With the ironing stage complete, allow the plastic to cool completely before using it.

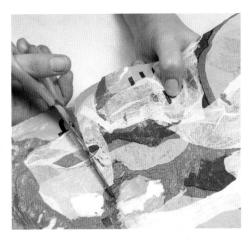

6 You can now use your plastic to make jewellery. Simply cut out your desired shapes using household scissors.

FINISHING TIPS

Once cool, fused plastic is surprisingly versatile. You can cut all sorts of shapes from it to make charms for bracelets or to layer up into intricate brooches or pendants. To make a plastic cuff, simply cut a rectangle of fused plastic. One dimension needs to match the circumference of your wrist; the other can be as deep as you want the cuff to be. Use scissors or a sharp knife to make holes in the short edges for attaching a fastening. Line up the edges so that holes at the top, middle and bottom on one edge have corresponding holes on the opposite edge. You can thread a length of ribbon through for tying or attach buttons on one edge and loops of elastic on the other to complete the piece.

Crocheting with plastic

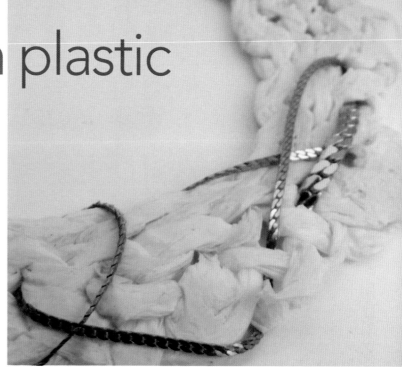

As with the sea string demonstrations on pages 34–37, you can apply crocheting techniques to plastic. Thin plastic bags, in particular, provide a fantastic material to work with. You can use them to create strong, yet lightweight, sculptural pieces. The example here shows how to make a base for a bib necklace, but you can easily use the same method to create a thick chunky cuff.

TOOLS AND MATERIALS:
- plastic bags
- scissors
- crochet hook
- iron
- clean sheet of paper

QUICK SIZE GUIDE

Because the plastic is quite bulky to work with, a large crochet hook is most suitable for this project, say 5mm. The strips of plastic you work with can measure anything from 5–10cm wide.

Crocheting a base

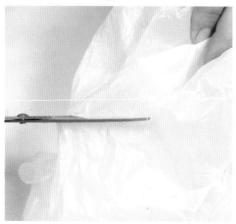

1 Begin by cutting your plastic bags into strips. It doesn't matter too much if they're not particularly straight or exactly even in width.

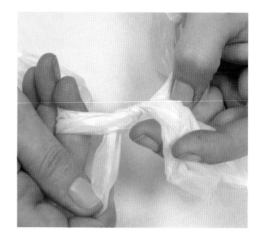

2 Tie the strips of plastic together to create one long continuous piece. Use double knots. You can trim any loose ends later on (see step 7).

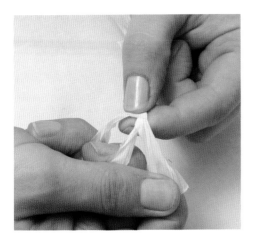

3 Starting at one end of your tied plastic strip, make a loop at one end and secure it with a knot. Again, use a double knot.

4 Put your crochet hook into the loop from front to back, bring the plastic round the hook anticlockwise and pull it through the loop.

WORKING A DESIGN

Making a cuff to go with this bib necklace is very easy. Simply follow the same steps, but skip any that drop stitches. A depth of around five rows is good for a chunky look. You want the cuff to be tight-fitting, so measure the circumference of your wrist and crochet to this length, making the fastening loop and bobble demonstrated on page 120 onto the main part of the cuff.

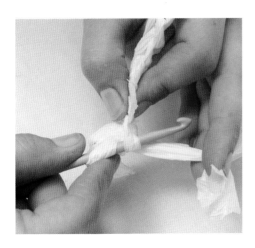

5 Follow steps 3 to 5 on page 35, to crochet a length of plastic chain around 20cm long. Remember to keep your work fairly loose.

6 When starting the second row (see steps 6 to 7 on page 35), skip going into the first chain. Dropping a stitch like this creates a curved shape.

7 Decide how many rows you want to work. In this example, three make a nice, deep crescent shape. Trim off any excess bits of plastic.

Crocheting a fastening

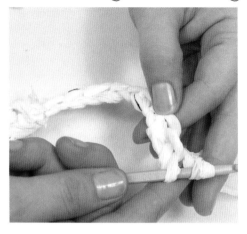

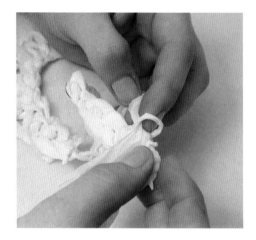

1 With the crescent complete, crochet a single chain from each corner to make a fastening. It needs to be around 15cm long on each side.

2 Crochet a few extra links on one length of chain. Make a loop at the end of the chain and tie a knot to secure it. Don't make the loop too big.

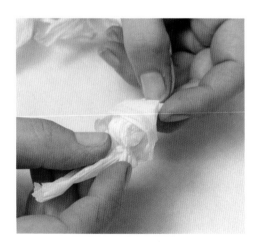

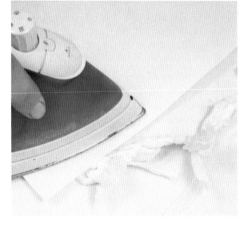

3 Turn to the other length of chain, and knot the plastic several times over itself to create a bobble that you can then push through the loop you made in step 2 to do the necklace up.

4 Arrange the finished piece on an ironing board and cover with a piece of paper. Fuse the plastic using a medium-hot iron with no steam. Remove the paper while warm, to prevent it sticking.

FINISHING TIPS

Once the ironed plastic piece has cooled sufficiently to handle, you can trim any untidy areas using sharp scissors. You can then decorate the base further, as here, with a length of chain.

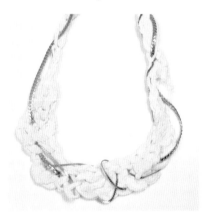

This example shows short pieces of found gold chain threaded through the loops of the plastic necklace. You can attach the chain to the back of your plastic piece using jump rings.

Making reflector jewels

Reflectors are great for adding a bit of sparkle to a piece of jewellery and work brilliantly stitched onto fabric using embroidery thread. A random arrangement works best.

TOOLS AND MATERIALS:
- broken bits of reflector (drilled and sanded, see page 111)
- piece of suede, leather or oilcloth
- craft knife
- ruler
- large, sharp needle
- embroidery thread

QUICK SIZE GUIDE

The size of fabric depends on your design. Use a 15 x 8cm piece for a cuff; a 15 x 2cm strip for a bracelet; or a 3–5cm diameter circle for a brooch or pendant.

FINISHING TIPS

To make a cuff, as here, sew one half of a press stud onto each short side. Place the 'female' piece on the right side of the fabric and the 'male' piece on the wrong side of the fabric, so that they are concealed when the piece is fastened.

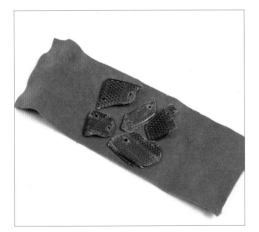

1 Cut your piece of suede or leather using a craft knife and ruler on a board. Place a number of reflector pieces on the suede until you have an arrangement that you like.

2 Stitch the pieces in place using embroidery thread (in a contrasting colour if you like). Go through the holes a couple of times as you do when sewing on buttons.

Making polymer clay beads

Clean and simple to use, polymer clay is a great material for making all manner of shapes into which you can set found things. The technique demonstrated on these pages is for making beads, but can also be used for making a pendant or focal piece on a bracelet. Look around for interesting embellishments like the nuts shown here – ring pulls from cans, bright-coloured metal bottle tops, tiny park pebbles – whatever takes your fancy.

TOOLS AND MATERIALS:
- polymer clay
- found metal pieces (bolts, nuts, ring-pulls)
- plastic knife or spatula
- rolling pin
- cocktail stick
- metal tray

QUICK SIZE GUIDE

For the earrings pictured, half of a 25-g pack of polymer clay was enough. You could use the same amount to make six to eight smaller beads for a bracelet or necklace, or a base for a striking pendant. Use half again for a focal bracelet embellishment.

1 Work the clay in your hands to make it soft, warm and pliable. Roll it out into a long sausage shape, around 15cm long. Cut the number of pieces you want, to the correct size, using a small plastic knife or spatula.

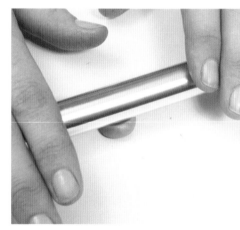

2 Roll each piece of clay into a neat ball. Then, using a small rolling pin, roll each piece out until it is about about ½cm thick. Don't overdo the rolling – you need the clay to be thick enough to press your found things into.

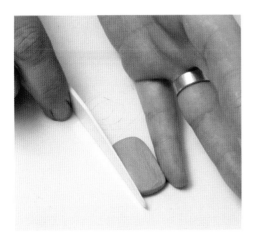

3 Give each bead some shape. You might want them all to be similar or more random – it is up to you. Use the knife to manipulate the clay and smooth the surfaces with your fingers.

4 Carefully press found things into each of your beads, taking care not to push them right through. Think about your design: you may want mirror image pieces for earrings, for example.

5 Don't move the found pieces about too much, as you want the clay to fit snugly and hold the pieces in place. Roll the surface of each bead to ensure a nice, flat finish.

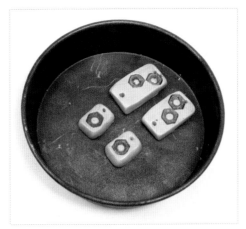

FINISHING TIPS

6 Plan how you'll join your beads together. Use a cocktail stick to poke one or two holes for connecting them with jump rings. This will depend on your intended design.

7 Place the beads on a metal tray and put them in a cool oven (140°C) for about 20 minutes. Take them out and allow to cool completely before incorporating them into a design.

You can square up polymer clay by filing or sanding the surfaces after baking. Connect pieces together using jump rings and finish with a hook and loop fastening for a necklace or bracelet, or attach earring wires.

Making a hose-clip ring shank

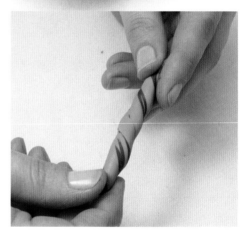

This basic technique demonstrates how to turn a found metal ring into a funky piece of wild jewellery using polymer clay. A hose clip is ideal for this: not only does its design provide a ready-to-use mount on which to set the clay, but it is also adjustable in size! A larger hose clip could be used to make a bracelet in the same way.

TOOLS AND MATERIALS:
- found hose clip
- polymer clay
- plastic knife or spatula
- baking tray

QUICK SIZE GUIDE

Polymer clay is available in 25-g packs in a wide range of colours. Experiment with different quantities, depending on the piece you are making: 5–10g is ample for a ring. The piece shown here uses twice as much of one colour as the other in order to achieve the marbling.

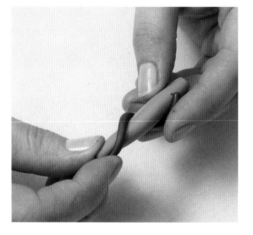

1 Using clay in two contrasting colours, work each piece until really soft and pliable. Roll each out, to about 6cm in length. Wrap the thinner roll around the outside of the fatter roll.

2 Gently twist the two colours together, so that they start to blend. This will give the finished piece a beautiful marbled effect. You could just as easily make the focal piece from one colour.

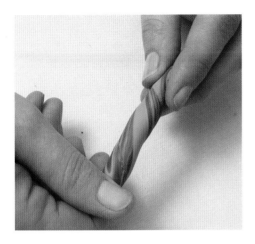

3 Keep twisting the clay, until the two colours are well combined. Don't overdo it, as you still want the colours to be distinct from one another.

4 Wrap the roll of twisted clay around the base of the hose clip. Use the clay to make a frame around the rusted metal at the centre of your design.

5 Use a plastic knife to trim the polymer clay where the edges overlap. Carefully smooth over the join with your fingers to meld the two ends together.

6 The shape of the finished piece can be quite free-form and organic looking, in which case simply manipulate it with your fingers. For a more regular shape, use a plastic knife or spatula.

7 Push the piece, clay-side down, onto your work surface. Apply pressure evenly to give the clay a flat face and to make sure that it sits flush with the metal of the hose clip.

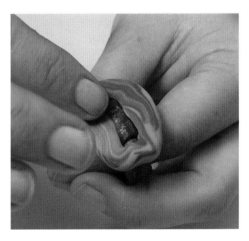

8 Rub the clay surface to smooth out any creases, before placing on a baking tray in a cool oven (140°C) for about 20 minutes. Allow the piece to cool completely before wearing.

Making resin beads

Crystal liquid epoxy resin is a useful medium for incorporating fragile finds into your jewellery designs – things that you might otherwise overlook, such as grass, leaves, berries and flowers from the city park. The resin sets hard around your found things and can be used to make linkable tablet beads and handmade cabochons for rings and pendants. Any moulds can be used.

QUICK SIZE GUIDE

Crystal liquid epoxy resin is usually supplied in two parts – hardener and resin – and mixed in quantities of 1 part hardener to 2 parts resin. Always read the manufacturer's instructions before handling, however. For this technique, 25ml of hardener to 50ml of resin was enough to make around ten flattish, beads between 1.5cm and 4cm long.

Making the beads

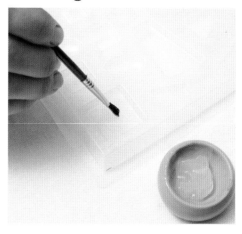

1 Work on a clean, dry surface and make sure you have all your tools and materials to hand, as the resin has a tendency to set quickly. Using a paintbrush, apply a thin coating of petroleum jelly to each of your moulds.

TOOLS AND MATERIALS:

- crystal liquid epoxy resin
- grass
- paintbrush
- petroleum jelly
- moulds
- latex gloves
- goggles
- two paper cups
- wooden skewer
- coarse sandpaper
- emery paper
- multi-tool with drill bit

2 Wearing gloves and goggles, pour the hardener and resin into a paper cup. Slowly mix them together. You usually get a wooden stick with the kit, but a lollipop stick is just as good.

3 Pour the mixture into a second cup and continue mixing carefully for a couple more minutes. This is an important part of the process, and will help prevent air bubbles forming.

4 Slowly pour the resin straight from the cup into your moulds. Pour steadily in one spot so that any air bubbles are forced out. Fill each of the moulds to about three-quarters full.

 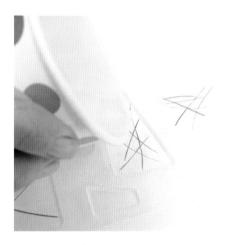 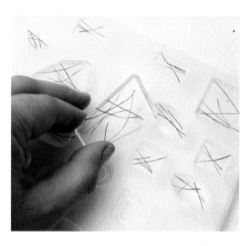

5 Arrange the things you want to set in the resin – in this case, grass. Push them into place with a wooden skewer. You do not have to be too precious about this – a random look is good and they might move about a bit anyway.

6 Pour a little more resin into each mould to cover your urban finds. Be careful not to pour too much liquid in. If it overlaps the edge of the mould you'll get a lip on your piece that can be difficult to get rid of.

7 Leave the resin to set in the moulds for at least 24 hours – check the manufacturer's instructions. Each of the pieces must be properly hardened before you can start working on them with sandpaper or drilling holes.

Preparing beads for use

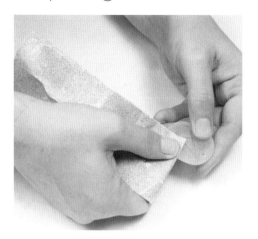

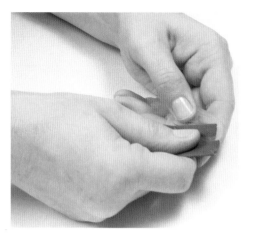

1 You have the option of two finishes: clear or opaque. For a clear finish, simply leave the pieces as they are. For an opaque finish, start by sanding each of the beads roughly using a coarse sandpaper.

2 Once you have sanded the entire surface, go over each piece again using a fine emery paper for a really smooth finish. Depending on the look you are after, you could even leave one side polished and one side opaque.

3 To prepare your resin beads for incorporating into a piece of jewellery, simply drill a hole at the top for a pendant or an earring drop, holes top and bottom for a necklace or at each of the sides for a bracelet.

FINISHING TIPS

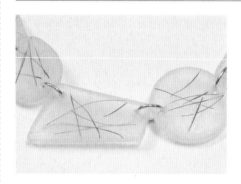

For a necklace or bracelet, link beads together using large jump rings. You can tie a ribbon or attach a ready-made chain to the final jump ring at each end to fasten the piece.

A single resin bead makes a striking embellishment for a statement ring. The simplest way of making it is to stick the resin piece to a shank using a hot-glue gun.

A bright-coloured sweet wrapper has been set into resin to make this eye-catching pendant. Simply attach to a ready-made chain using a jump ring for an instant necklace.

Making a claw setting

This is a great technique for showcasing the pretty, irregular shapes of many city-park pebbles. A simple silver claw setting can be adapted to make a statement ring embellishment, or a delicate chain of linked pebbles for a bracelet. The setting is minimalist in design, allowing you to admire the qualities of the stones themselves.

TOOLS AND MATERIALS:
- flattish pebble
- 1.5-mm gauge silver wire
- 0.8-mm gauge silver wire
- wire cutters
- flat-nosed pliers
- snips
- silver solder
- metal plate
- hammer
- paintbrush
- borax cone
- goggles
- soldering torch
- plastic tweezers
- pickle
- fireproof brick
- metal file

QUICK SIZE GUIDE

A setting for a 1–2cm-diameter pebble can be made using 10cm of 1.5-mm gauge silver wire for the foundation, and 20cm of 0.8-mm gauge wire for the claws. (These need to be around 2cm long, depending on the size of your pebble). You can make several settings and connect them together for a bracelet or make a ring by soldering a setting onto a 2-mm wire ring shank before setting the stone.

Preparing the setting

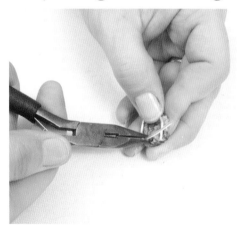

1 Cut a length of 1.5-mm gauge wire that will fit the diameter of your pebble. Use flat-nosed pliers to shape the wire: it needs to sit beneath it, so should be a little bit smaller than the pebble.

CONTINUED

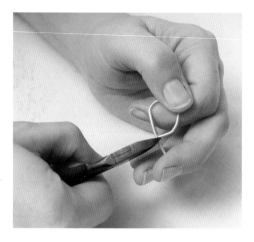

2 Snip off any excess wire and solder the ends of the loop together. Wearing goggles, follow steps 3–7 on page 45 to prepare the solder, apply borax and heat the link.

3 After soldering the link, place the wire loop in pickle to cool. Then remove, rinse and dry it. Hammer the piece gently on a metal plate to flatten it out a bit and give it a textured finish.

4 Prepare the solder for the claws. Cut six lengths of 0.8-mm gauge wire. Working on a fireproof brick, arrange the claws at regular intervals on the wire loop, applying borax and solder first.

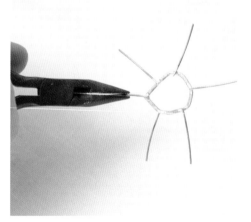

5 Wearing goggles, gently heat the setting using a low torch flame. Take care not to blow the solder or the silver claws out of position.

6 Use a steady, circular motion to heat the piece evenly and stop when the solder starts to melt. Lift the piece with flat-nosed pliers and place in pickle.

7 After pickling, rinse and dry the setting. Then, carefully, hammer the soldered areas and silver claws a little to flatten them.

wild jewellery

Setting the stone

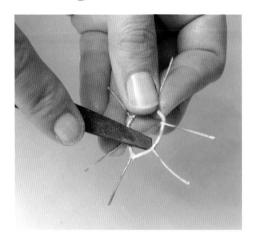

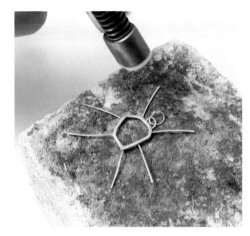

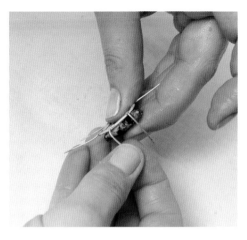

1 Use a metal file to smooth out any rough edges. Take extra care here, as you do not want to break any claws off. Make two small jump rings out of 0.8-mm gauge wire.

2 Solder the first jump ring onto the back of the claw frame. Hook the second jump ring onto the first and solder that one closed. This provides a link with which to attach a chain.

3 Position your pebble and hold firmly as you start to bend the claws around the pebble using your fingers. Bend each claw gradually so that you get a good fit for the setting.

4 Trim each claw a little using snips: each one should bend up the side of the pebble and onto the front.

5 File the end of each claw to smooth out any sharp points that might catch on fabrics or fingers when worn.

6 Use flat-nosed pliers to squeeze the claws in more so they are really snug around the sides of the pebble.

Urban inspiration

In many ways, urban finds present the greatest challenge when making wild jewellery. Discarded plastic, dull cardboard and rusty nuts and bolts all seem a little uninspiring at first glance. And yet, a look through the techniques in this section of the book proves that the most innovative pieces can be made using these cast-off materials.

As with wild treasures from other environments, such as the beach or a woodland setting, the secret to success lies in making the most of the inherent characteristics a material might have – the melting properties of a plastic bag or the sparkle of foil wrapper, for example. You may discover that the wild elements of your jewellery designs rarely resemble the materials with which you started out. As such, urban inspiration may require a little experimentation, but will rarely fail to disappoint.

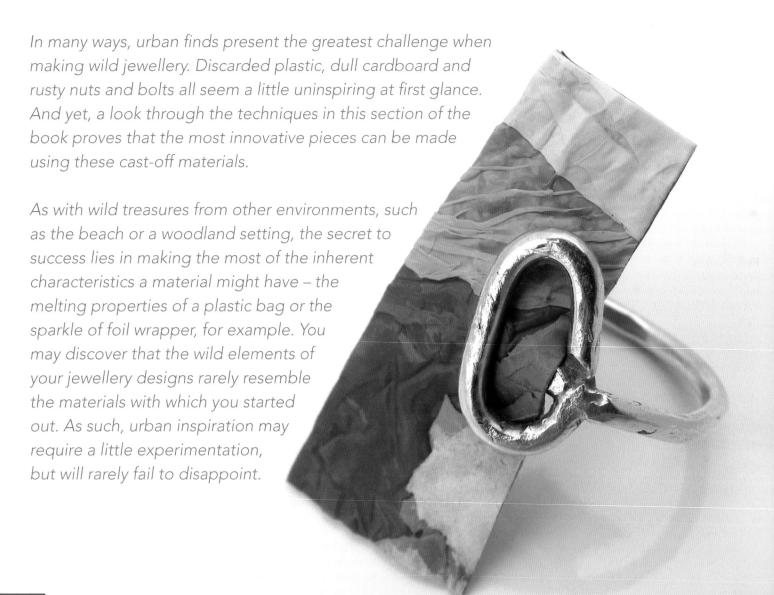

Spiral pendant
A single strip of corrugated card rolled into a snail-like spiral serves as a pendant.

Pendant reflector
A reflector fragment makes a sparkly pendant. (See page 111 for drilling plastic.)

Pink sparkle
Foil wrapper and resin bead ring ornament. (See pages 126–128 for resin beads.)

Bright vision
A red bead and reflector ring. (See pages 30–31 for making a wire ring shank.)

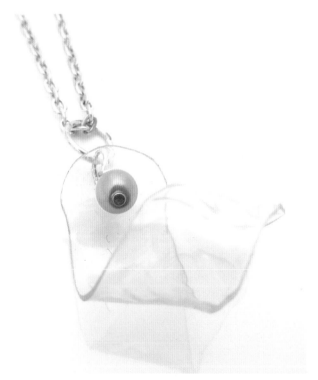

Jewellery by Emma Northcott

When it comes to making quirky, contemporary pieces, Emma Northcott from Emus Jewellery has found many ingenious uses for urban materials. Working from her studio in Cornwall, Emma collects and stores unwanted plastics, papers and metals for future use. She transforms mundane, everyday items into lovely, affordable, wearable jewellery.

Among Emma's fascinating creations are sculptural pendants made from plastic bottles, colourful bracelets with jigsaw-piece charms and abstract cuffs made from fused plastic bags. Perhaps most appealing about her work is that people are often left guessing as to exactly how a piece has been made, or from what. That, and the sheer versatility of her work, is testimony to Emma's extraordinary vision and design skills.

Below: A charm bracelet made from varnished jigsaw pieces.

Clockwise from top left:
A bracelet made using 'petals' cut from fused plastic bags.
A flower brooch carved from the inner tube of a car tyre.
A fused-plastic-bag cuff.
A curled plastic-bottle pendant.

Resources

There are two things to consider when making your own jewellery. The first is your personal style: what types of jewellery do you like to wear? What colour schemes appeal to you? Are you are fan of bold, chunky statement pieces, or do you prefer a certain understated elegance? The second thing to consider is your materials: how will your wild finds feature in a piece? How can you make the most of their natural shapes, colours or textures? What other things could you incorporate to make your finds sparkle? The following pages offer advice on how to answer some of these questions. You will also find a number of useful addresses for sourcing tools and materials.

Designing jewellery

The idea of designing your own jewellery can be a bit daunting at first. Some people think that they have to be really good at drawing to make beautiful pieces, but this is not really true. There is no mystery to the design process. In fact, if you have ever made anything at all, you're probably proficient at designing already, you just haven't realized it.

Designing is all about making decisions about the way things look. So if you pick a contrasting colour thread to sew on a fabric patch, decide to grate some beetroot on your salad for a bit of colour, or choose to wear red lipstick instead of pink, you're making a conscious decision aesthetically to change the look of something you're making or how you look.

The kinds of decision you make when putting designs together for jewellery-making are no different. The colour of the sea plastic you choose to put with some pearls you might have, the shade of twigs you go for to make a cuff, or the magazine page you opt for when rolling up beads – they all fit in with a vision you already have in your head. This could be a subconscious idea or a conscious, definite aim. You might have a vague feeling for the type of look you're after, that's sometimes difficult to put into words; you might have a well-thought-out plan, or have been given specific instructions for a particular piece. Whatever your motivation or inspiration, there are useful, practical ways in which you can organize your working methods or allow the more vague ideas to free themselves.

Seeking inspiration

It's a good idea to have a little notebook with you at all times so that you can quickly sketch things out or make lists of ideas as they come to you. Make an effort to collect bits and pieces that appeal to you whenever you see them: scraps of pretty fabric; motifs that you come across on packaging or wrapping paper; fragments of crockery with an interesting colour scheme. All of these things influence you one way or another. Kept almost as a scrapbook on a pinboard, in a file or pasted into an album, they provide a constant source of ideas to refer to time and again.

Allow yourself to take inspiration from other things that you are interested in – you will be surprised how they can find a way of being expressed in the pieces you make. If you are interested in modern and vintage fashion, look at magazines and books for current trends and colours that might translate well into a piece you have in mind. The same can be said for the colours and shapes you might find if you are interested in architecture or abstract paintings. Or perhaps you have a more vague sense of a look or a person you can imagine wearing your jewellery that might give you inspiration: a favourite singer, a stylish friend. It is in this way that you can make jewellery that is truly original in design, unique in execution and satisfying in construction. Once you allow yourself to be creative with your designs, inspiration can come from almost any source.

Work in progress

Using found materials to make jewellery really gives designing a helping hand and keeps pieces fresh and original. You never know exactly what you're going to find, and right from the moment you bend down to pick something up you've begun the design process, just by choosing an item that catches your eye for its colour, shape or texture. It is so much easier and enjoyable to think about things you'd like to make when you're outside and out and about, choosing unusual, surprising materials. Then, when you take your stash of found treasure back home to make into lovely things, putting a colour scheme or design together is easier because you already have your found treasures as a starting point.

It is good practice to keep complementary bead and button selections together in jars or on saucers. You can then readily select them to lay them out with found materials to check how colours and textures work together when they are next to each other. Laying selected pieces out in the basic shape of a necklace or bracelet, for example, will help you to decide which shapes work well next to each other. Have to hand a wide selection of ribbons, cord and chain for threading and mounting your finds. Making your own fastenings can be fun (see Making cold connections, pages 47 and 52–53), but you can just as easily buy them ready made (see Useful addresses, pages 142–143). This way you have the choice of making something on the spot or labouring on a piece. And don't be afraid to draw things. You may end up with strange, squiggly sketches, but they can very often be a welcome starting point and just the trigger you need to make the most of the materials you have to hand.

wild jewellery

Useful addresses

Below is a list of jewellery suppliers with shops that you can visit for tools, wire, findings and beads. Many of these businesses can also be visited and bought from via the internet. In recent years, the internet has made it so much easier to find materials and findings from a wide range of sources and, because there are now so many sites, the prices are very competitive. As well as listing the shops, therefore, there is also a handful of websites that come highly recommended. Ebay is great for ordering specific items, such as tigertail and pearls, often with really low postage. It is also good for second-hand tools.

Of course, depending on where you live, you might have a local craft, bead or fabric shop where you can find materials for making jewellery. Sometimes it is nice to choose items by handling them, looking closely at the colours and thinking about your designs. And don't rule out local markets and car-boot sales for cheap tools. DIY shops and department stores are useful – and sometimes cheaper than specialist suppliers and websites – for other tools and materials you might need, such as a multi-tool, saw and sandpaper.

UNITED KINGDOM

Tools

J Blundell & Sons
25 St Cross Street
London EC1N 8UH
Tel: +44 (0207) 437 4746
www.jblundells.uk.com

Buck & Ryan
Shop 4, Southampton Row
London WC1B 4DA
Tel: +44 (0207) 636 7475
www.buckandryan.co.uk

Le Ronka
Unit 3 Sandy Lane
Stourport-on-Severn
Worcestershire
DY13 9PT
Tel: +44 (01299) 873 600
www.leronka.co.uk

Wire and metals

Cookson Precious Metals Ltd
59–83 Vittoria Street
Birmingham
B1 3NZ
Tel: +44 (0121) 200 2120
www.cooksongold.com

Findings

Exchange Findings
49 Hatton Garden
London EC1N 8YS
Tel: +44 (0207) 831 7574

Samuel Findings & Jewellers Ltd
14 St Cross Street
London EC1N 8UN
Tel. +44 (0207) 831 0657

H A Light Findings Ltd
The Rical Group
Tramway
Oldbury Road
Smethwick
West Midlands
B66 1NY
Tel: +44 (0121) 555 8395
www.lightfindings.co.uk

T H Findings
42 Hylton Street
Birmingham
B18 6HN
Tel: +44 (0121) 554 9889
www.thfindings.com

Beads and gemstones

Capital Gems
30B Great Sutton Street
London EC1V 0DU
Tel: +44 (0207) 253 3575
www.capitalgems.com

R Holt & Co
98 Hatton Garden
London EC1N 8NX
Tel: +44 (0207) 430 5284
www.holtsgems.com

Levy Gems Ltd
26-27 Minerva House
Hatton Garden
London EC1N 8BR
Tel: +44 (0207) 242 4547
www.levygems.com

Manchester Minerals
Georges Road
Stockport
Cheshire
SK4 1DP
Tel: +44 (0161) 477 0435
www.manchesterminerals.co.uk

wild jewellery

Websites

www.beads.co.uk
Wide selection of wire, beads
and findings; very good value

www.cooksongold.com
Good prices for metal, silver wire
and tools; no minimum order

www.wires.co.uk
Good for large quantities of wire
in different sizes and finishes

www.ebay.co.uk
Good for specific materials,
findings and cheap tools

www.bigbadbeads.co.uk
Well-priced feature beads to
off-set your found items

UNITED STATES

Tools

Allcraft Tool & Supply Co
135 West 29th Street #402
New York, NY 10001
Tel: +1 (800) 645 7124

Anchor Tool & Supply Inc
PO Box 265
Chatham, NJ 07928
Tel: +1 (201) 887 8888

Armstrong Tool & Supply Co
31747 West Eight Mile Road
Livonia, MI 48152
Tel: +1 (800) 446 9694
www.armstrongtool.com

Frei & Borel
PO Box 796
126 Second Street
Oakland, CA 94604
Tel: +1 (510) 832 0355
www.ofrei.com

Indian Jeweler's Supply
601 East Coal Avenue
Gallup, NM 87305
Tel: +1 (505) 772 4451
www.ijsinc.com

Wire and metals

David H Fell & Company
6009 Bandini Blvd
City of Commerce, CA 90040
Tel: +1 (323) 722 6567
www.dhfco.com

T B Hagstoz & Son
709 Sansom Street
Philadelphia, PA 19106
Tel: +1 (215) 922 1627
www.hagstoz.com

Handy & Harman
Camden Metals
12244 Willow Grove Road
Camden, DE 19934
Tel: +1 (302) 697 9521
www.handytube.com

Hauser & Miller Co
10950 Lin-Valle Drive
St Louis, MO 63123
Tel: +1 (800) 462 7447
www.hauserandmiller.com

Belden Wire and Cable
Company
PO Box 1327
350 NW N Street
Richmond, IN 47374
Tel: +1 (765) 962 7561
www.belden.com

Findings

Halstead Bead Inc
6650 Inter-Cal Way
Prescott, AZ 8630
Tel: +1 (800) 528 0535
www.halsteadbead.com

New York Findings
72 Bowery
New York NY 10013
Tel: +1 (888) 925 5745
www.newyorkfindings.com

Rings & Things
304 E 2nd Ave
Spokane, WA 99202
Tel: +1 (800) 366 2156
www.rings-things.com

Roseco
13740 Omega Road
Dallas, TX 75244
Tel: +1 (800) 527 4490
www.roseco.com

Beads and gemstones

Du Lyon Jewelers
202 Walton Way
Suite 192-229
Cedar Park, TX 78613
Tel: +1 (512) 4702036
www.dulyon.com

Paraiba International
East 46th Street, Suite 1100
New York, NY 10017
Tel: +1 (877) 888 1080
www.paraibainternational.com

Fire Mountain Gems
1 Fire Mountain Way
Grants Pass, OR 97526–2373
Tel: +1 800 355 2137
www.firemountaingems.com

CANADA

Tools

1400 Ages Drive
Ottawa
ON K1B 4K9
Tel: +1 (613) 526 4695
www.busybeetools.com

Lacy and Co Ltd
69 Queen Street East
Toronto
ON M5C 1R6
Tel: +1 (416) 365 1375
www.lacytools.com

Wire and metals

Imperial Smelting & Refining
Co Ltd
451 Denison
Markham
ON L3R 1B7
Tel: +1 (905) 475 9566
www.imperialproducts.com

Johnson Matthey Ltd
130 Gliddon Road
Brampton
ON L6W 3M8
Tel: +1 (905) 453 6120
www.matthey.com

AUSTRALIA

Wire and metals

A & E Metal Merchants
68 Smith Street
Marrickville
NSW 2204
Tel: +61 (2) 8568 4200
www.aemetal.com.au

Johnson Matthey
64 Lillie Crescent
Tullamarine
Melbourne
VIC 3043
Tel: +61 (3) 9344 770064
www.matthey.com

Index